Feeling the Meaning of Life

How the Evolution of the Senses Led to Asking the Question

Yves Sauvé, PhD

For more information, email ysauve@ualberta.ca

Names: Sauvé, Yves, 1964- author
Title: Feeling the Meaning of Life: How the Evolution of the Senses Led to Asking the Question / Yves Sauvé. Includes bibliographical references and index.
Identifiers: ISBN 978-1-77136-952-7
Subjects: LCSH: Brain [QP376-QP430 (Physiology)]; Evolution [QH359-QH425 (Biology)]; Vision [QP474-QP495 (Physiology)]; Sensorimotor integration [QP454].
Set in Garamond, UK grammar
Designed by Yves Sauvé

Contents

"*What I'm writing to you is not for reading; it's for being.*"
Clarice Lispector (Água Viva)

To Doris and Etienne

Reader's Rights

You can write in this book if it's yours to rightfully own. If you borrowed it from a library or a friend, think twice before altering its physicality; a question of compassion, which we all have the potential to exert. Borrowing this book might put less pressure on your feeling that you should read it all. You might want to only read the prologue or perhaps jump to specific sections, or simply browse through the figures. If you have an electronic version, you can search for specific words; may I recommend "consciousness", "mental models", or "compassion"?

This book puts forward that the soul arises from mental models of our internal and external world. That is, neurons are involved in modelling likely futures and planning optimal actions for our survival and reproduction. We evolved from single cells 3.8 billion years ago precisely because our myriad of ancestors survived and exchanged genetic information. You can search for any words you'd like. These are your rights as a reader.

According to author Daniel Pennac (Comme un Roman), readers have the right to: "*1) not read; 2) skip pages, skip whole sections; 3) not finish reading the book (perhaps you feel you've read something like that before, or you disagree with the author's point of view or simply the writing style, the book style, it might get on your nerves); 4) read it again, for fun, not skipping pages or sections, adopting a different perspective, revisiting specific points; 5) read whatever you feel like, even works that you think are repetitive, or stereotypic, or are the insipid fruits of lucrative purposes; 6) get physically excited, undergo visceral reactions, be contaminated by textually transmitted diseases; 7) reading anywhere, like the soldier who volunteers to clean toilets so he can read the complete works of Nicolas Gogol; 8) glance, jump at any places in the book, if you just have very limited time to read; 9) read aloud, like this little girl who explained she loved reading out loud because her school had prohibited it, she joins many authors who do re-read their work aloud; and 10) be silent, to not share your own experience and feelings vis-à-vis a book*". With the ubiquitous smartphones, Pennac added an 11th right, the right to turn them off, to read peacefully. He concluded his list by reminding us to never make fun of the person who does not currently read, as this person is a potential reader of your own writings!

You will consult this book, using your senses.

Now that your rights have been laid out, please use them. There will be, at times, intricate details provided as a reminder of what is known. This knowledge is factual: it is part of the story. These seemingly rebarbative details are a glimpse, an example, and a humbling experience of what can be learned by the curious.

As enlightenment philosopher Immanuel Kant put it: "*Have the courage to use your own understanding*". Without such an approach, one is at risk of being distracted, misled, and even taken advantage of by proponents of truth. Kant had a firm stance: "*Laziness and cowardice are the reasons why such a large part of mankind gladly remain minors all their lives, long after nature has freed them from external guidance. They are the reasons why it is so easy for others to set themselves up as guardians. It is so comfortable to be a minor*".

Prologue

The reality your brain constructs is not accurate, but instead limited to a narrow range of physical events; in the case of reading this book, it is two-dimensional black on white shapes. Try reading this:

7H15 M3554G3 53RV35 7O PR0V3
H0W 0UR M1ND5 C4N D0 4M4Z1NG 7H1NG5!
1MPR3551V3 7H1NG5!
1N 7H3 B3G1NN1NG 17 W45 H4RD BU7
N0W, 0N 7H15 LIN3
Y0UR M1ND 1S R34D1NG 17
4U70M471C4LLY W17H0U7 3V3N 7H1NK1NG 4B0U7 17
Y0U 5H0ULD B3 4W4RE 0P WH47 15 R34L 4ND WH47 15 N07
W3 4RE W1R3D 70 H4V3 4 B14553D V13W 0P 7H3 W0RLD

If you can read the paragraph above, and it makes sense, then you are showing a bias. The manner in which you take in the world around you is a biased one. You are not detecting the shapes, the letters and numbers you have learned as individual symbols; your brain creates the most familiar pattern based on your past experience, conscious or not. Your eyes merely modulate the production of the brain into what you perceive as real. The first part of this book will take us through the steps of how we see, from light entering our eyes, up to the point that we can consciously state: I see. I see the light at the end of the tunnel, and hopefully it is not a train. If it is, then I must move quickly, because I intend to live. There is so much to learn.

So you D1D R34D the text with capital letters and numbers, intertwined, and you made sense of nonsense, approaching it as a game. Now consider all that you have learned, including English or other languages. Any variations from what you are used to reading and hearing are transformed, in order to look like what you are used to reading and hearing, based on your very own personal experience. These "shortcuts", also fancily called "heuristics", can induce social and interpersonal biases. Such biases apply to modalities other than verbal language… body language, appearances, speech, social stances, political views, belief systems, even your own sense of self. But how do we "perceive"? What is "perceiving"? And while we are at it, how do we "think"? What is, "thinking"? Do other animals perceive and think? Can single cells, including bacteria, think and perceive? Let us "think" about it.

Here I am digressing; how easy it is to lose track. We were talking about perception and somehow got to talk about thoughts. But hang on, because questioning whether single cells perceive and think is relevant to understanding how we, *Homo sapiens*, perceive and think. Well, let me make it relevant by asking even more questions. Where do we come from? What is life? How did life start? I admit that I am digressing yet again, and the book has not even begun officially. So yes we can do it, we can perceive and we can think. Our ancestors did too, perceive and think. How far can we go with our ancestors?

We are here, you are here, you exist, and you are reading this book, because single cells evolved to become multicellular organisms. What? Well, yes, your story can be traced to single cells, dating 3.8 billion years ago. But at which point were your ancestors able to perceive and think? How about we examine how we have evolved from single cells to become *Homo sapiens*? More precisely, as a means to understand how we perceive and think, I propose that we look at how perception and thoughts have evolved.

Under the pressure for multicellular organisms to survive and reproduce, nervous systems have evolved to generate a reality which entails sources of threats and pro-survival (food). Reproduction involves passing on information in the form of genes between and across generations. We will explore the multiple requirements for both survival and reproduction. The experience of reality is generated by the nervous system and can be modulated by sensory inputs (internal or external), or without any inputs (thoughts: ruminations, unconscious occurrences that profusely seep into our consciousness), The ultimate purpose is to detect threats, be they real or perceived, and to achieve reproduction of our species until genes can be passed on to another generation.

We will take the time to review what we know about evolution: both the vertical transmission from one organism to its descendants, but also the horizontal transmission between species through other means of gene exchange, such as carried by bacteriophages or other viruses, and through cells incorporating bacterial mitochondria (genetic symbiosis). There are plenty of bacteriophages around us. They represent the most abundant and various life forms. Their population is estimated to reach a number equivalent to ten followed by 30 zeros; and mitochondria, while containing only 30 genes, possess many more copies than the nucleus does. The bottom line is this: the evidence we have gathered together as a society supports that we evolved from single cells dating 3.8 billion years ago, which required the survival and reproduction of species (vertical exchange) and genetic information exchange between species (horizontal exchange), until you were born, yes you the reader, you the curious reader, receptive to learning with an open mind. You might feel like smiling, thinking "me, I am the outcome of 3.8 billion years of evolution?" You might feel grateful and say: I would like to thank my parents, and my grand parents, and all my ancestors. Cutting to the chase: I gratefulness for these first single cells, and the unique conditions and ample time that allowed them to form.

As the cultural anthropologist Mary Catherine Bateson said, "*We are not what we know but what we are willing to learn.*" So, be proud of "being". It is my honour to participate in your willingness to learn. I am just a messenger... and an expert at digression. That said, these digressions have a purpose: to contagiously inspire you to observe the reality we are discovering by making unexpected links. The word "science" in the context of this book is synonymous to "observing" and "taking in" the world. As such, with the ultimate goal of presenting facts, I will minimize the use of specialized jargon, but there will be also sections with intricate details, which you have the right as the reader to simply skip over.

The point is that the information will be made available as an option to accommodate all types of readers. Facts will be presented with the context that interconnects them. Again, the word "science" is not meant to imply "a private club for members only", but rather to honour our efforts to observe the reality we have evolved in, using our biological senses and the technology we have developed and to continue developing as a society that grasps these facts.

The fact that the brain itself contains no internal sensory pathways means that thoughts occur within what appears to be a void. The generation of conscious perceptions does not entail generating a perception of the neurons generating this perception; otherwise, it would represent an infinite loop. In other words, we are not conscious of the processes that generate thoughts. The feeling of thoughts does not represent the generation of thoughts. Thought processes, while based in physiology (facts), bear no resemblance to sensory experiences. Sensing is the generation of a reality driven by the evolutionary pressure to survive and reproduce. To the unaware, this phenomenon can be both mysterious and threatening. We are threat detectors; across evolution, threat detection has ensured survival of an organism until reproduction could be achieved, so that genes could be passed on. As threat detectors, we humans have responded by adopting various extensive and often deeply emotionally rooted belief systems. A natural reflex has been the distinction between the mind and the physical brain, and this is known as dualism or non-reductionism.

History and present day abound with examples of candidates acting as external agents: as the site of the mind, as the sacred, as the soul. We have a powerful urge to believe. Even more so, this is what our brain does. Our brain evolved to create predictive mental models of the physical world we interact with (internal and external to us). This very function of our brain makes it possible for us to create such external agents; they are models of how we conceive the world, of how we respond to the apparent threat of not "feeling" the processes generating these very models and these thoughts. Scientific advances also offer more candidates as external agents, such as within the fields of quantum mechanics and new physics. We will carefully explore how mental models evolved, starting from our ultimate ancestors, single cells dating 3.8 billion years ago, to the anthropocentric *Homo sapiens.* Ultimately, the continuous process of evolution has led to a nervous system which generates the reality that we experience. The operating word here is "continuous"; what we call the soul did not appear at a specific point through a mysterious event. Our reality is limited to the physical modalities that had to be detected so as to ensure the survival and reproduction of our ancestors, and is equally limited to the technology we are developing.

We lack the imagination to be humbled by how limited our representation is of the physical world we live in. The time and spatial scales we can comprehend are determined by our senses and the technology we develop… we do not "see" atoms, we do not feel the distortion of time by gravitational forces… we do not "see" any of the 4 forces we have identified… we know them because of the tools we have developed: 1- electromagnetic, 2- gravitational, 3- weak nuclear force and 4- strong nuclear forces. This is just a language limited by our senses and technological innovations. Together, space and time form an infinite continuum, with no beginning, and no end.

This book proposes a reflection that does not restrict us to being materialistic but represents an opportunity to express compassion for each other: an opportunity to survive and therefore evolve as a whole.

You can reflect on how you are contributing to our society, on how you are also dependent on the role and expertise of others: remember the person cleaning the floor, the person emptying the garbage, the factory workers, the diaper changers, the educators, the countless individuals in the shadow doing so many intricate tasks. We need each other… we very, very much do, equally.

Two of the hurdles to continuing our evolution are:

I. Ignorance, from which stems unfounded, irrational fears at every level. As Derek Bok put it: "*If you think education is expensive, try ignorance*".

II. Convenience. Why bother unlearning all emotionally deeply rooted beliefs we were taught as part of our cultural upbringing, to simply observe reality using our physical senses with deliberate non-bias and with uncompromised compassion? Why bother letting go of long cultivated habits? Why bother being fully present in the present moment, with acceptance and without judgment? Why? To be enlightened, as Kant meant it.

We are carrying genes. Once we accept it, we are free to live, we are no longer slaves of the mechanisms, drives, and forces that allow genes to change (evolve). We can evolve as a society, one breath at a time. We become aware that thoughts are deceitful. We are free of Plato's myth of the cave. I smile because I am happy, but I can simply be happy because I smile.

"*The soul in which philosophy dwells should by its health make even the body healthy. It should make its tranquillity and gladness shine out from within; should form in its own mould the outward demeanour, and consequently arm it with a graceful pride, an active and joyous bearing, and a contented and good-natured countenance. The surest sign of wisdom is constant cheerfulness.*" Michel de Montaigne (Essays)

Chapter I

Vision

"*The eye sees only what the mind is prepared to comprehend.*"
Henri Louis Bergson

The book you are now reading with your eyes, or fingers using braille, or listening to as an audiobook, or that your dog is smelling or tasting, is an opportunity to stop and reflect on the evolution of the senses, on how your senses relate to the reality you experience, the reality that your brain generates, creates, builds up, emulates, constructs, produces. Visual illusions provide evidence that you "see" with your brain, that your eyes (retinas) modulate the construction of reality by your brain.

The same holds for illusions based on other senses. Let us consider an illusion inspired by Akiyoshi Kitaoka (figure 1), which depicts two circles. It is hard to conceive that you are looking at two perfect circles. You can use a round glass, a cup, or any other circular object(s) to verify that these are indeed two perfect circles. If you do this with a cup of coffee, the stains will be evidence enough. Your brain is fooled. What is happening here? Many events take place in your occipital lobe. If you make deliberate efforts to render the image blurry, you will better appreciate the roundness of both circles; there is no longer an illusion when the edges are blurred.

Figure 1

CIRCULAR ARGUMENTS ARE HIDDEN

BEHIND TWISTED VIEWS OF REALITY

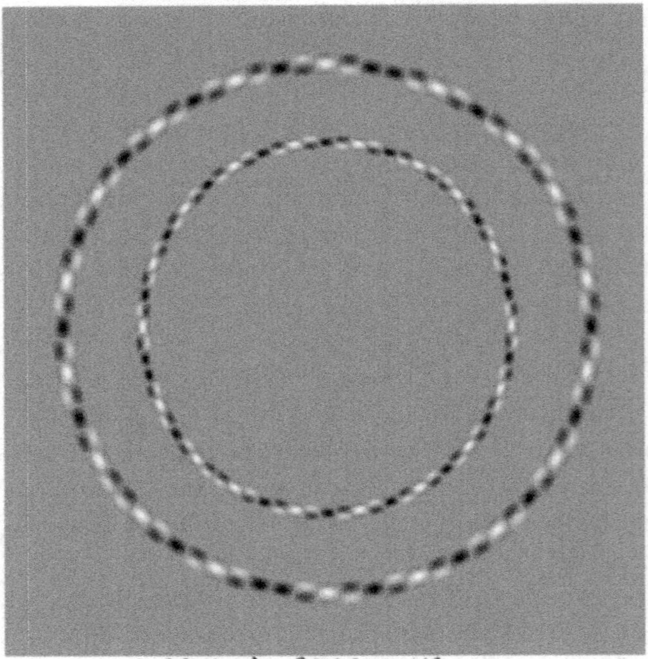

... CIRCULAR REASONING
WORKS BECAUSE CIRCULAR REASONING
WORKS BECAUSE CIRCULAR REASONING WORKS
BECAUSE CIRCULAR REASONING WORKS BECAUSE...

Have a go, be an experimenter. You can close your eyes until you can barely see the image. If you have reading glasses try removing them, or if you do not have reading glasses try standing as far away as you can while still seeing the circles. You can draw one conclusion from your experiment: the illusion involves the perception (conscious or not) of multiple individual black and white rectangles of specific location and orientation. Your brain links the array of distinct rectangles (edges, contrasts, transitions of bright-dark, technically speaking: Gabor patches), creating a distorted ensemble. When deliberately not focusing on the image as part of your experiment, the inability for individual rectangles to form a focused image on your central retina precludes seeing the illusion. There is more information to be gathered here, looking at each Gabor patch individually, also precludes experiencing the illusion. You can use a computer or magnifying glass to zoom in on individual rectangles to test this phenomenon.

Hence, what we see is conditional and contextual. The brain generates what we see; it builds an image based on distinct cues. When you see a face, your brain constructs the image of this very unique face by linking an array of individual Gabor-like patches, each with their respective contrast, position and orientation.

In the process of looking at a face, your eyes jump from one Gabor-like patch to the next, producing what is known as microsaccades (Shelchkova et al., 2019). In this manner, your eyes automatically peruse the contour of the face and of its key features such as eyes, nose, mouth and facial hair (Hosoya and

Hyvärinen, 2017). Jumps can be as short in duration as 20-30 milliseconds and as small as 15 minutes of arc (15 divisions out of 60 within a single angular degree), a distance corresponding to the length of this hyphen - if you are looking at it at arms length. For familiar faces, the whole process can be as fast as 180 milliseconds (Visconti di Oleggio Castello and Gobbini 2015). Not only do you generate such micro-jumps when looking at faces, saccades are part of how your brain constructs your experience of the visual world around you.

Take your eyes away from your book and look around you. You are producing numerous saccades. If you are at a cocktail party and like me, you do not know what to say, ask someone to do this exercise with you: join your hands together over your head, forming a circle with your arms, ask this someone to peruse the contour of the circle. You will be observing saccades. Feel free to take turns; the ice has been melted, enjoy your cocktail. You can share with your interlocutor that each saccade lasts about 0.1 second, during which there is no conscious vision. You are blind every time you generate a saccade (saccadic masking or suppression). A window of opportunity to steal a drink? Well, you would have to be quick and synchronize your theft exactly with the 0.1 second eye jumps. An opportunity for dextrous magicians perhaps? (Rieiro et al., 2013).

If you tally the total time spent doing saccades every waking day (about 150,000), then you get over 4 hours each day during which our visual perception does not accurately reflect what is right in front of our eyes (Melcher and Colby, 2008). You can challenge your cocktail partner by decreeing that we are blind on an average of 4 hours every day. He or she might become more nervous (a good sign) and start blinking more often, and so might you, if you care. More time added to being blind. The average person blinks 17,000 to 22,000 times in a day, each blink lasting 0.1 second, with a total time spent blind estimated from 1,700 to 2,200 seconds, or 28 to 37 minutes. The total, therefore, approximates 4.5 hours of being blind every waking day. More icebreaking, more drinks ahead. Has your interlocutor started to divide into two interlocutors? Excessive alcohol intake is infamous for seeing double (diplopia); another proof that the images you see are a construct of your brain. How far are you ready to push experimenting?

The linkage or binding of separate edges (contrasts) is elemental to how the brain generates images. Furthering our understanding of how we fall for specific illusions, means learning more about how we actually see. Beware of con-text. Here is another example, known as the café wall illusion, drawing its roots from an 1898 study by A.H. Pierce and his observations of analogous patterns in kindergarten weaving place mats (figure 2). Again, the mechanisms involve distinct edges-patches-borders of specific location, orientation, size and contrast, each modulating cellular activity in the visual cortex (Takeuchi, 2005).

I sense that you might be able to see the building blocks that your brain uses to generate the images you experience as reality. Do you realize what I mean? It looks like hard work, but glance again. Picture this: you are bright, luminous, and perceptive. Trust my vision of things to foresee. I am looking forward to further illustrate how marvellous your senses are. Envisage all the possibilities. Together, we will see it; visual illusions are real (figures 3 and 4).

Figure 2

Figure 3

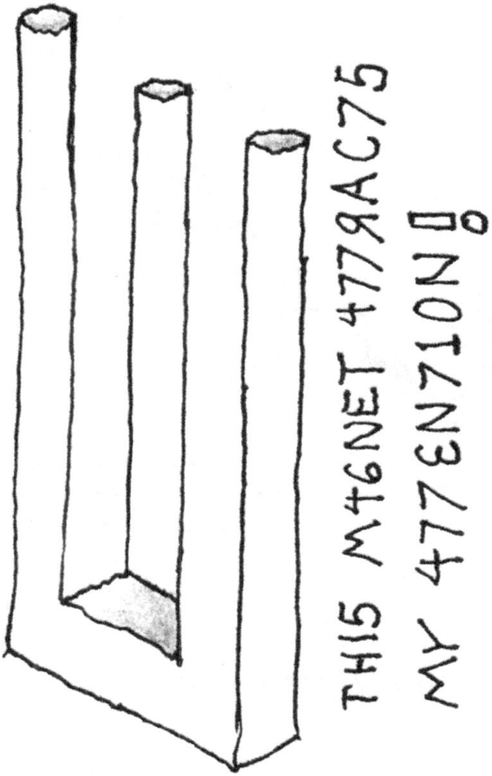

What does it mean to see? I see with my eyes, but I absolutely require my brain to perceive or detect what my eyes "see". I believe what I "see"… But most importantly, I see what I believe, in this order. What determines what I see is my brain. The eye modulates brain activity, but does not generate the image I see. Ultimately, I see with my brain. My brain generates the reality I experience (and this is true for all senses).

Figure 4

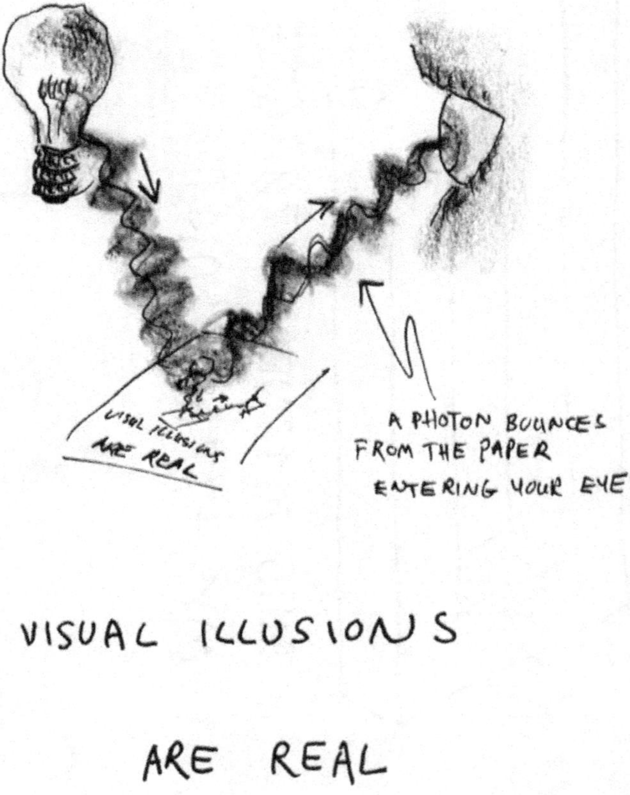

What do we know about the actual events that take place when we use our eyes to see the world around us, such as the illusion depicted in figure 3? If the illusion is an image on a paper, then photons (actually many, many, of them, around a billion...) will bounce on the paper and converge on the outer part of your fovea (figure 5).

Figure 5

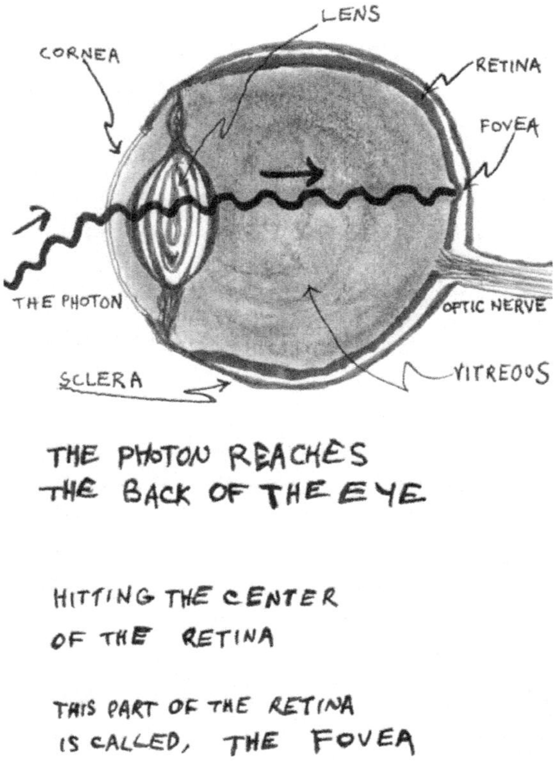

If the source is a light-emitting monitor, photons will directly bombard your fovea, so long as you are looking at the illusion. Looking at it means to unconsciously enabling the photons to converge in your retina, precisely on the outer segments of your cone photoreceptors where they are most concentrated: the fovea (figure 6).

But how does this occur? Three unconscious events (reflexes) must take place for this first step to be successful: 1) your pupils will constrict; 2) your eyes will converge toward your nose (the closer the image, the more crossed your eyes will be; ask a friend to look at the tip of your finger as you bring it closer to her-his face); 3) the lens inside your eyes will become more curved, rounder, and as you get older this bending becomes limited (this is why you tend to pull the paper, or smartphone, away from you to read and see it). In addition, as is the case with facial recognition, you will be doing microsaccades when looking at the illusion.

Photons are a form of energy. They are both particles and waves, but not at the same time, at least not when studied with our current scientific tools. When studied as particles, photons exhibit many quantifiable properties such as energy, but "size" does not apply to them. When studied as waves, photons have a wavelength. In human photoreceptors, retinal (a photosensitive vitamin A derivative) is covalently bound (strong chemical bound) with a protein, opsin (from the greek ὄψις, opsis, for "sight"). Visible light, in human, varies from 460 to 720 nanometres. The smaller the wavelength, the more energy it carries.

Let us focus on one photon, out of over a billion in daylight, caught by the outer segments of a cone photoreceptor. Here, it is a green cone. There are three types of cones in the human retina, red, green and blue. Only green and red are found in the fovea, the area of the retina responsible for maximal visual acuity; figure 6). The photon is absorbed by a molecule called a chromophore (figure 7); "chromo": light, "phore": carrier. The chromophore is "11-*cis*-

retinal", but more commonly you will hear the term "retinal".

If we further magnify the cone outer segment, we can see a protein called green opsin that spans the whole membrane, binding the infamous chromophore in its core with the strongest chemical bonds: the "covalent bounds" (figure 8). Not even 1/1000 second (millisecond) has passed, and the photon will now cease to exist in its current form. Light energy is absorbed by a double bound between two carbon atoms (figure 9), unlocking retinal, allowing it to flip (isomerize) from its crooked *cis* isomer to its straight *trans* isomer (figure 10). The chromophore is now locked into the *trans* isomer shape, and is now straight and will stay as such; changing it back to the *cis* form will require a complex series of processes known as the retinoid cycle, but this is for another time.

Exchanging one form of energy to another form is called transduction. All events in sensing involve converting one form of energy (stimulus, input) to another (reaction, output). Transduction is the ultimate foundation of all forms of sensing, internal and external to organisms. Transduction is an elemental aspect of all life forms. All cells, whether they are standalone species (unicellular) or part of an organism (multicellular), are characterized by their ability to detect a signal, and to transform this signal into a cellular event. The nature of the signal and of the ensuing cellular event varies between cells.

Figure 6

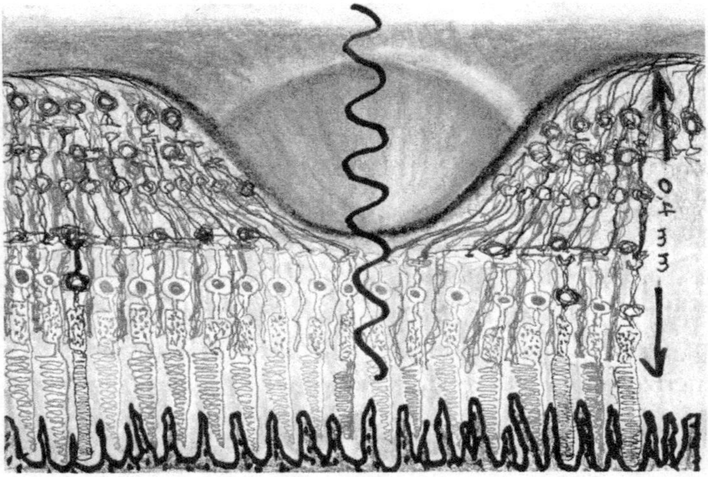

Figure 7

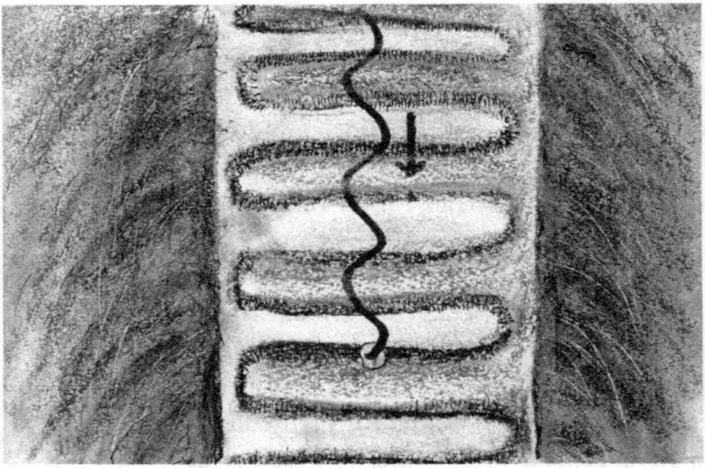

Figure 8

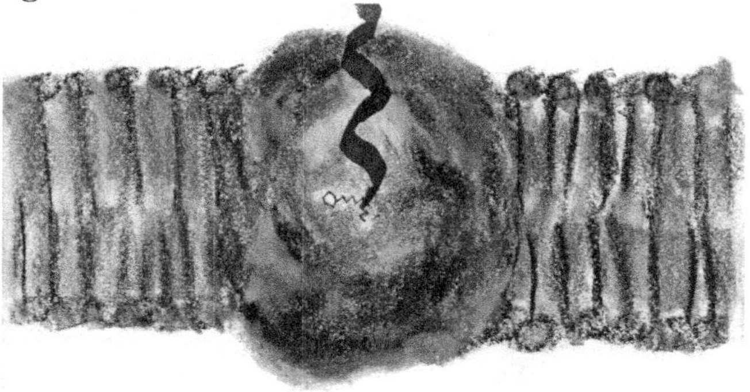

Figure 9

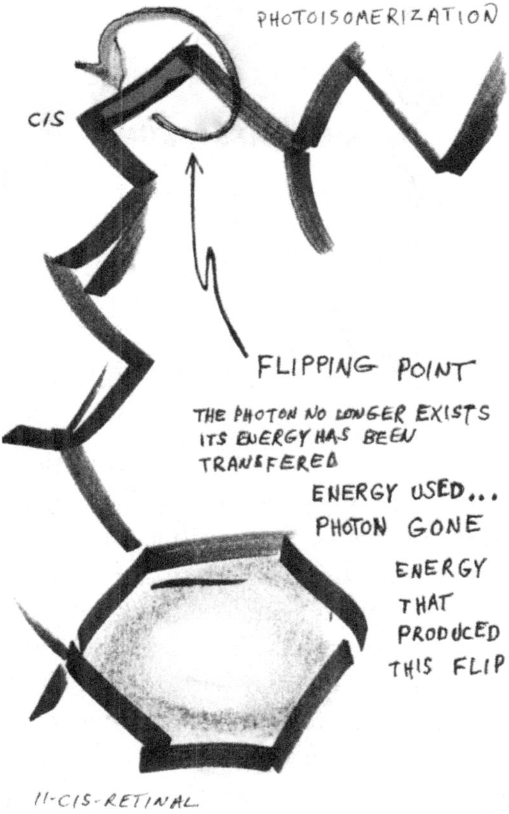

Figure 10

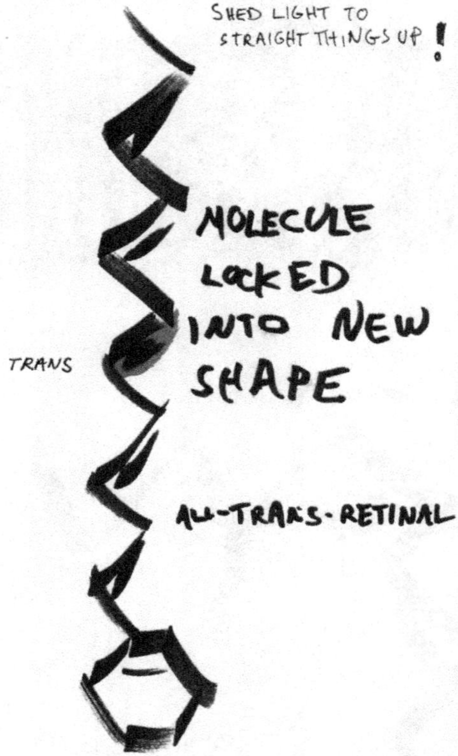

Sensing light is achieved through phototransduction. If you think about it, photo stands for photon (light) and isomerization implies converting one isomer of the chromophore retinal, *cis* (giving a crooked shape) to another form, *trans* (giving a straight shape). At this very time, light has no role to play in seeing. What will now follow is a series of events that are absolutely independent of light... we are "feeling" that what we shall see will be a product of all these light-independent events... many of them... seeing is so much more than catching light (photons). Indeed, at this point, once the chromophore retinal is straightened up using the energy of the captured photon (figure 11), light has nothing to do with vision. You have not seen anything yet... literally; prepare yourself for a wild ride. All is a matter of perception (figure 12). As is the case with all aspects of conscious vision, the only visual effect itself is a shape change in a small molecule.

Figure 11

Figure 12

AS IT IS THE CASE WITH ANY ASPECTS OF CONSCIOUS VISION, THE ONLY "VISUAL" EFFECT (LIGHT ITSELF) OF VISUAL ILLUSIONS ON YOU (YOUR BODY) IS THE SHAPE CHANGE OF A SMALL MOLECULE, 11-CIS-RETINAL, INTO ALL-TRANS-RETINAL ...
THE MANY OTHER EVENTS (BIOCHEMICAL, ELECTRICAL), OCCUR WITHOUT YOUR AWARENESS
YOUR BRAIN EMULATES REALITY AS YOU PERCEIVE IT, SOMETIMES... IT MAKES MISTAKES!!!

IT IS ALL A MATTER OF PERCEPTION

An ode to transduction, as all life forms depend on it for their survival and gene transmission. Transduction is our link with reality, including our own physiology. The reality we experience relies on transduction; our senses are transducers, they transform an input (stimulus) into an output (reaction). Etymology of "feel", from "*Old English felan to touch or have a sensory experience of; perceive, sense (something),*" in late Old English "*have a mental perception...*". The connecting notion might be "*perceive through senses which are not referred to any special organ.*" (https://www.etymonline.com/word/Feel).

"*All nature has a feeling: woods, fields, brooks*
Are life eternal: and in silence they
Speak happiness beyond the reach of books;
There's nothing mortal in them; their decay
Is the green life of change; to pass away
And come again in blooms revivified.
Its birth was heaven, eternal it its stay,
And with the sun and moon shall still abide
Beneath their day and night and heaven wide."

John Clare

Exercise of re-interpreting John Clare's poem in the context that sensing is, at its very foundation, transducing.

We connect with reality by transducing specific forms energy to build a representation of woods, fields and brooks. These are light (photons bouncing from woods, fields and brooks), mechanical (sound waves, contact of skin, scent and taste derived from wood, fields and brooks). I form a mental construct, which stands alone, becomes an entity, which now exists and now persists independently of nature, therefore becoming eternal. In silence, and it could be also without any other sensory stimuli, (in darkness, without scent, touch) they speak, because now there is a mental representation, without the need of information from books, movies, music, painting or any other media.... Once my brain has generated the construct, the image, the representation, the emotions linked with it, the woods, fields and brooks are immortal as far as I am concerned. Each time I recall this moment, it is recreated from multiple aspects, imperfectly, without my awareness. I am fooled. I have the illusion of mentally revisiting the same exact seasons, in an eternal loop of seeming perfection. I can imagine my mental model, reconstructed again and again at every recall, in the context of the moon and the sun, in fact in relation to any other contextual elements, in any forms that transcend the notions of night and day and that epitomize my own concept of heaven, limitlessly.

Figure 13

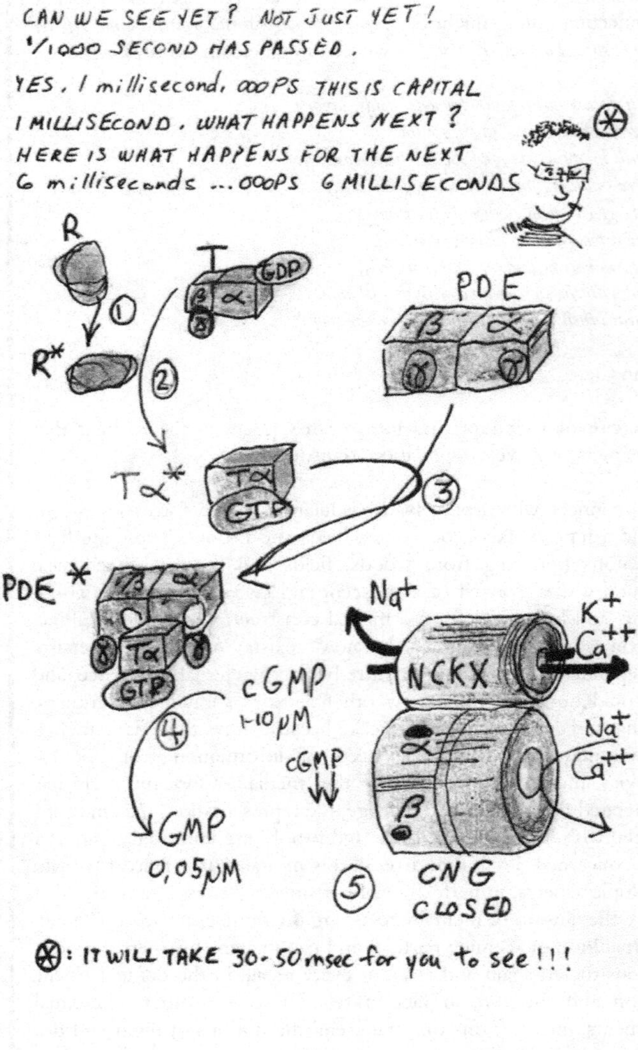
CAN WE SEE YET? NOT JUST YET!
1/1000 SECOND HAS PASSED.
YES, 1 millisecond, OOOPS THIS IS CAPITAL
1 MILLISECOND. WHAT HAPPENS NEXT?
HERE IS WHAT HAPPENS FOR THE NEXT
6 milliseconds ...OOOPS 6 MILLISECONDS
R
1
R*
T
GDP
β
α
γ
2
PDE
β
α
γ
γ
Tα*
Tα
GTP
3
PDE*
β
α
γ
Tα
γ
GTP
Na+
K+
Ca++
NCKX
cGMP
1-10μM
4
cGMP
α
Na+
Ca++
β
GMP
0,05μM
5
CNG
CLOSED
⊛: IT WILL TAKE 30-50 msec for you to see!!!

There are so many steps involved, so many opportunities for variations (as in errors), all of which contribute to the image perceived.

One millisecond (1/1000 second) has passed. The next events, described in figure 14, will culminate after the next 6 milliseconds in the stoppage of a current known as the "dark-current". The voltage (electrical potential, positive charge content) will be reduced inside the photoreceptor. In a nutshell, the more there are molecules changing shapes, the greater the reduction in positive charges will be inside the photoreceptor.

Lets get started. To begin with, picture that the chromophore is tightly bound (covalent bound, high energy chemical link) with a large protein (rhodopsin, R). The implication is that the light-induced shape change of the chromophore (*trans-*) modifies the shape of rhodopsin (step 1), which is now activated (R*). Rhodopsin is a member of a class of membrane bound receptors called G protein-coupled receptors. When activated, a G-protein called transducin (T) is able to bind to R* (step 2), like a key fitting in a lock. Now bound to R*, T looses two of its proteins (beta and gamma), keeping just the alpha. Transducin-alpha is able to receive energy from the reaction of guanosine diphosphate (GDP) becoming guanosine triphosphate (GTP). Transducin-alpha is now activated (Tα*) and can bind to another key protein complex, phosphodiesterase (PDE). The binding of Tα* to PDE causes it to lose two alpha proteins, turning this complex (PDE*, step 3) into an enzyme; that is, PDE facilitates a chemical reaction. Here, this reaction is the conversion of cGMP (cyclic guanosine monophosphate) into GMP (step 4). Less cGMP is available to bind CNG channel (cGMP-gated nucleotide gated) that allows positive ions (sodium and calcium) entering the photoreceptor (step 5).

The flow of positive ions inside photoreceptors is called "dark-current", because it flows when uninterrupted by phototransduction. Closure of the CNG channel stops the dark-current, at least in part, because there are many more CNG channels in photoreceptors (figure 14). With the assistance of an ion exchanger (NCKX, sodium, calcium, potassium exchanger, not shown on figure 14), the amount of net positive charges inside the photoreceptor declines. Therefore, the more photons are caught by the photoreceptor, the greater the effect on stopping the dark-current will be.

We can talk about the notion of threshold here: to elicit a visual response in the brain, a critical level of dark-current stoppage is required. This minimum will depend on the quantity and duration of previous light exposure, on the type of photoreceptor and on the type of responses elicited, such as causing visual reflexes or producing a conscious percept by the brain that could entail detecting light level changes or object movements, or of actual images including illusions.

Seeing the illusions will depend on stopping a critical amount of dark current; this is non-negotiable. Thus, the photoreceptors that will entice your visual experience must already generate a dark-current; in daylight, rod photoreceptors do not. Saturated by photons, rods are inactivated; they are unresponsive to visual stimuli. Removing the flooding of rods by the daylight rain of photons (photopic adaptation) will allow rods to start generating the dark-current. A

Figure 14

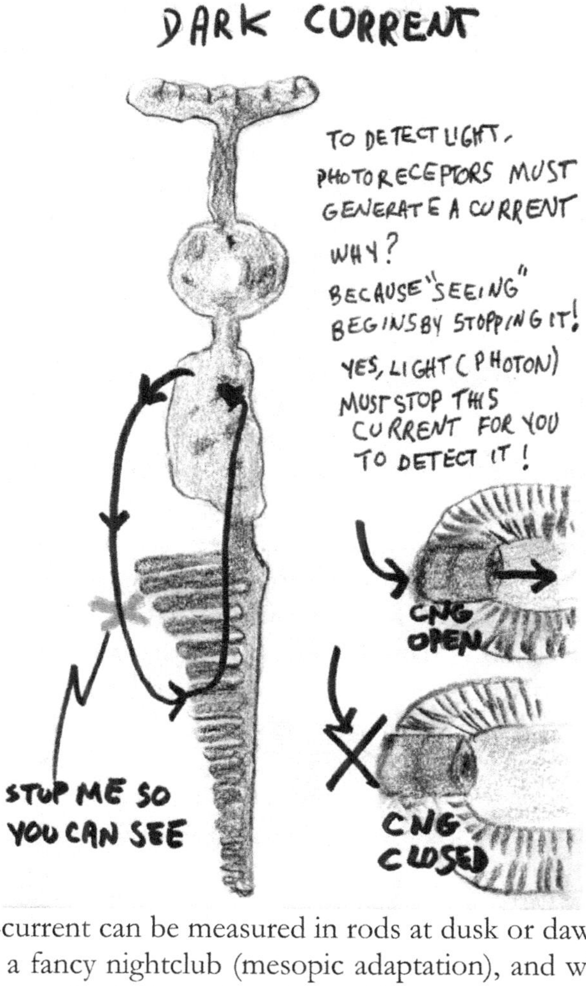

dark-current can be measured in rods at dusk or dawn or in a fancy nightclub (mesopic adaptation), and will reach maximal flow after about 20 minutes in total darkness (scotopic adaptation).

Cones have the unique property of being able to still generate a dark-current despite torrential photons raining, that is, even on a sunny day at the beach. Despite the fact that, thinking about it, the term "dark-current" is a misnomer, can you spot a contradiction here? The events depicted in figure 13 consider rhodopsin, which is the type of photopigment (opsin family) expressed by rods. How can this figure explain the first step in seeing the illusions? It is your cones, not your rods that are involved here in seeing the illusions. Well, here is an issue that takes us back to how phototransduction was studied. Scientists relied heavily on rats and mice, which are nocturnal animals, with 98% of photoreceptors being rods. Thus, we know much more about the physiology of rods than cones. What if the most common laboratory animals had had cone-dominated retinas, such as ground squirrels, chameleons, or fishes, would we know more about daylight vision? Not quite. Surprisingly, the population of cones in the human retina does not even exceed 5% of all photoreceptors. Our retinas are made up of over 95% rods, and yet we have evolved to be active during daylight. We are diurnal animals, relying mostly on a dense packing of our cones in our central retina, the fovea. In a way, one can say that when it comes to seeing fine details, we have tunnel vision. A look back at evolution adds context to a seemingly paradoxical constitution of our retinas.

Here I am digressing, so let us get back to the next steps.

About 7 milliseconds have passed since we looked at the illusions, the dark-current has stopped in the cones that underwent phototransduction, exceeding thresholds. This will impact the electrical activity of other cells in the retina (figure 15): cone bipolar cells (BC) and horizontal cells (H).

Figure 15

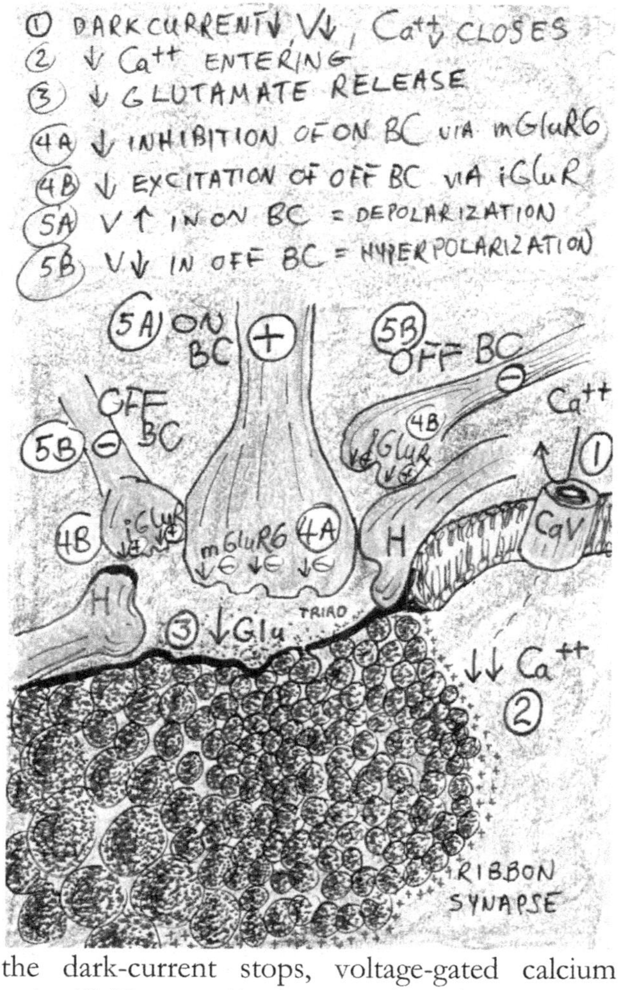

As the dark-current stops, voltage-gated calcium channels (CaV, step 1) close at the photoreceptor axon terminal. Less positively charged calcium ions enter the ribbon synapse, meaning less charge repulsion against the positively charged synaptic vesicles containing glutamate (step 2). Fewer vesicles release neurotransmitters (glutamate) in the synaptic space (step 3). Less glutamate binds to its receptors (step 4).

The conventional wisdom is that glutamate acts as a neurotransmitter, particularly of the excitatory type. In the retina, glutamate can be both: excitatory or inhibitory. In fact, what determines the action of any neurotransmitter (or any molecule binding to its receptor) is the nature of the receptor itself.

Both types of glutamate receptors, excitatory (ionotropic) and inhibitory (metabotropic), are found in the retina. Cones establish contact with cells (bipolar cells), which bear the inhibitory (ON BC) or the excitatory (OFF BC) receptor type on the other side of the synapse (post-synaptic side). The photoreceptor's response to light always results in stopping glutamate release, diminishing the amount that binds to its receptors. A diminution in glutamate will remove the post-synaptic inhibition on ON BC, exciting them (step 5A), while it will stop exciting OFF BC, inhibiting them (step 5B).

These events are the same across all species possessing photoreceptors. A key feature of all the synapses established by photoreceptors is that they contact at least three post-synaptic neurons (one bipolar cell and two horizontal cells), which led to the name "triad". Cone bipolar cells establish excitatory synapses with retinal ganglion cells (glutamate binding on excitatory post-synaptic receptors located on RGCs). If RGCs become positive enough, they will generate action potentials; these are the very and only signals sent to the brain, through the one million RGC axons that together bundle up to form the optic nerve (figure 16).

It will have taken a minimum of about 30 milliseconds for the first information to be sent through the optic nerve. What type of information does the retina send to the brain? We went over the light-driven stoppage of a current in photoreceptors, which through a first synapse excited or inhibited bipolar cell activity, and then through a second synaptic level either excited or inhibited retinal ganglion cells. Each bipolar cell receives synaptic input from photoreceptor clusters segregated into a central and a peripheral region (centre surround receptive fields). The clusters are the smallest in the fovea, the region of the retina with which we fixate (foveate); as such, synaptic convergence is at its lowest in the fovea, and accordingly, so are the receptive field sizes. Horizontal cells (H, figure 15) mediate the inhibition between neighbouring central and peripheral clusters, so that if photons are simultaneously transduced in photoreceptors from both clusters, the outcome is no modulation of activity in that bipolar cell. In this manner, light only modulates bipolar cell activity when it falls at selective regions of its photoreceptor clusters. This configuration leads to the detection of edges. These edges represent the basic information about shape discrimination sent by the retina to the brain: black and white edges, as richly contained in individual Gabor patches.

Look again at the illusions in figures 1 and 2; observe the "edges" and reflect on the role they play in your experience of the illusions. The visual illusions, can we "see" them now?

No, not yet. Remember: we see with our brain (cortex); our retina merely modulates its generation of reality. We *sense* with our eyes, but see with our brain.

Figure 16

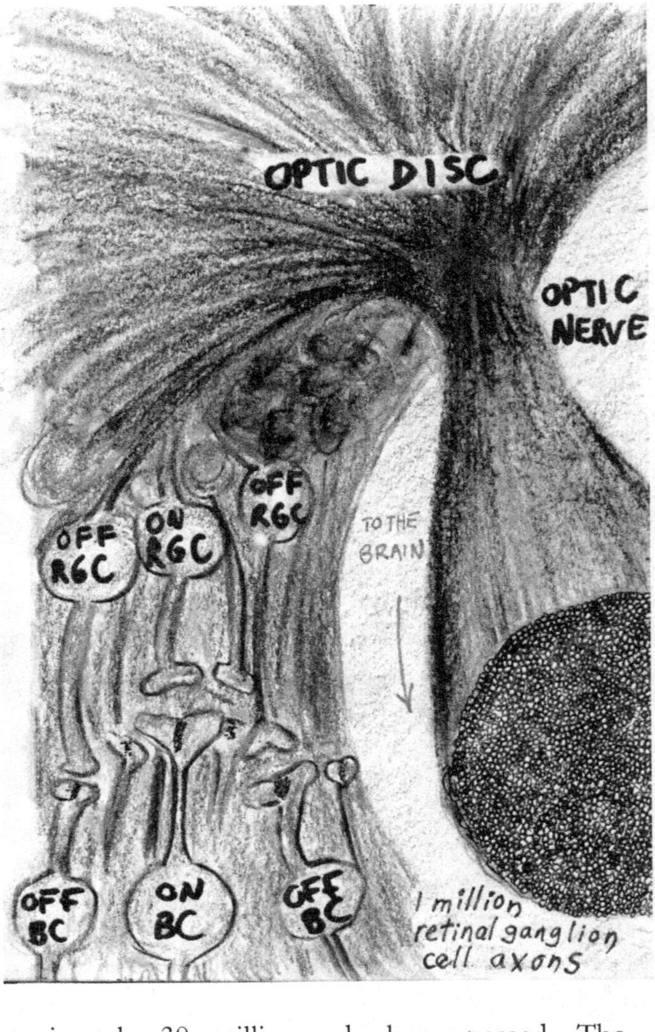

Approximately 30 milliseconds have passed. The information about the illusions is leaving the retina, carried through solely as nerve impulses (action potentials) in individual RGCs, projecting their axons through the optic nerve. None of them go directly to the cortex. Zilch.

Figure 17

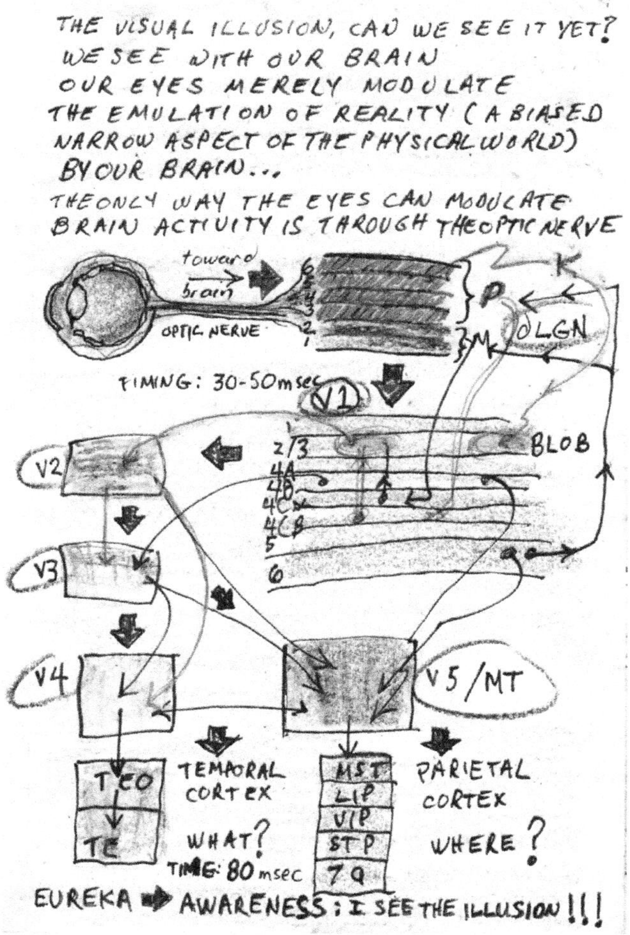

I am sorry to disappoint you, to make you linger, but the retina does not modulate the cortex, not directly. Conscious vision is mediated by RGCs projecting axons to a part of a subcortical region called the thalamus, inside a nucleus (a subdivision of the thalamus) named the dorsolateral geniculate nucleus (dLGN). Information about specific aspects of the visual scenes is sent along three segregated pathways (figure 17):

1) The parvocellular pathway: Information about fine details, shapes, object identification, conveyed by small size RGCs (P-cells, midget) projecting axons establishing synapses on dLGN relay neurons in layers 3-6, which in turn send axons to layer 4 of the primary visual cortex (V1), then to V2, V3, V4, followed by multiple regions of the temporal cortex.

2) The magnocellular pathway: Information about movement and large shapes, biased toward more peripheral vision conveyed by larger size RGCs (M-cells, parasol), projecting axons synapsing on dLGN relay neurons in layers 1 and 2, which in turn send axons to layer 4 of V1, then to V2, V3, V5 followed by multiple regions of the parietal cortex.

Finally, 3) the koniocellular pathway: information about colour, conveyed by middle size RGCs (K cells, blue-yellow) projecting axons synapsing on dLGN relay neurons located between each of the six layers of the dLGN, which in turn send axons directly to layers 2-3 of the primary visual cortex (V1), in clusters known as "blobs", which in turn send information to V2 and V3. A minimum of 80 milliseconds is required for cortical areas to become modulated after looking at the illusions, but you only experience it by about 400 milliseconds; just less than half a second.

Rather than acting as a camera transferring an intact image to the brain, the retina deconstructs the image (that was focused on its surface, as indeed a camera) into distinct features that are analysed by anatomically and functionally distinct pathways in the brain, building the visual reality we experience. In addition to the three pathways mentioned above (parvocellular for fine details, magnocellular for motion, koniocellular for colour), image creation requires two more key features: location of centre-surround receptive fields in the retina (two-dimensional aspect) and computation of disparity in receptive fields location between both eyes (three-dimensional aspect, depth perception).

To appreciate what disparity means and is, look at this square □ with one eye at a time (the square will change location, moving right when changing from right to left eye, and moving back left when switching back to the right eye). The illusion of depth perception in 3-D movies is created by the projection of two movies, to be looked at in each of your respective eyes. Glasses with polarized lenses (vertical and horizontal) have replaced the use of red and green lenses to drive the segregated stimulation of the right versus left eyes. When you are watching a 3-D movie, each eye is stimulated by a different movie with the only difference being the disparity between specific objects in the scene. The greater the disparity, the closer the object is perceived. For instance, a fast approaching comet is perceived as passing nearby when its image falls at remote corresponding areas of your left and right retinas, just as is the case when looking at the square above. Try the exercise again. Look at this □ occluding one eye at a time. Now back off, so you can look as far away as possible and repeat the exercise; the square will not change position when alternating between eyes. Perhaps you can look at something on a wall in front of you, or at someone in a café (please note that the author declines responsibilities for any potential reactions if you use blinking to switch from looking with the left and right eye). The outcome is: the image does not move. A mountain is perceived as distant in a 3-D movie when its image falls exactly at the same place on each retina. For this reason, having crossed eyes precludes experiencing 3-D movies.

When you are looking, information from independently analysed features is ultimately combined by cross-talk between as up to 50 different cortical areas, building what you see. Instead of analysing one feature at a time, as occurs in most computers (serial processors), the simultaneous processing of distinct features saves precious time. But still, we see in the past. The time taken for the brain to build a visual percept can vary from 80 to 400 milliseconds, depending on the type of stimuli (movement, abstract patterns, facial expressions, illusions) and the related output task (motor or cognitive); Eklund and Wiens, 2018; Franklin et al., 2016; Sato et al., 2015. What we see has occurred in the past, after this time window has passed.

It takes time for your brain to construct the image you experience. Looking at our evolution, acting as threat detectors, had to be able to process life-compromising stimuli and act accordingly to avoid the threat. A similar concept applies for life-promoting stimuli such as when hunting for food. Let us consider tripping over a fallen branch or rock and having to change our posture to protect ourselves from getting injured. Our predictive mental models have evolved to predict the collision course in a manner permitting action in a timely fashion, such as covering our face with our arms. How do we know this? More illusions: the Hering illusion (Figure 18), in which you see both vertical lines as curving outward in the centre.

Figure 18

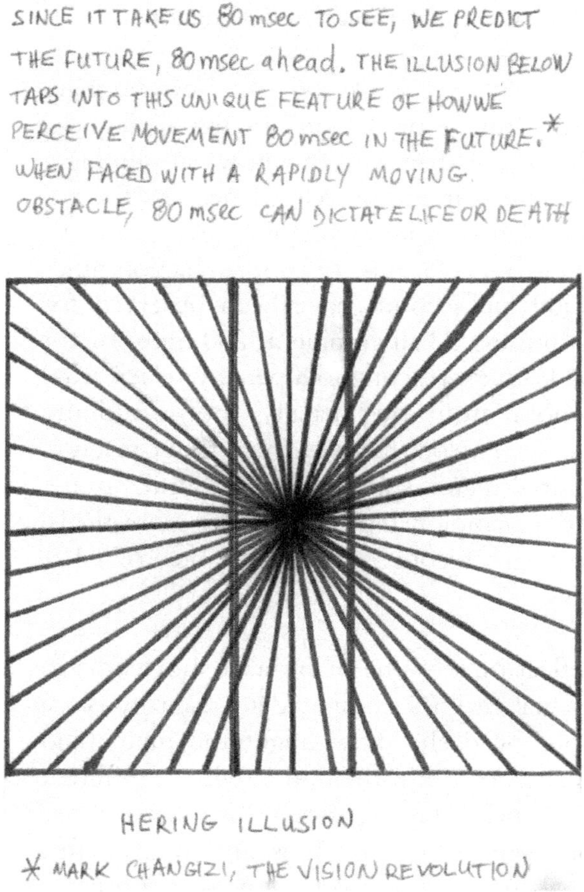

Mark Changizi put forward the "perceiving the present" hypothesis, stipulating that the visual system mitigates processing delays by spatially warping scenes into what they should appear as approximately 100 ms in the future (Changizi, 2001). When you look at the Hering illusion, you are tricked into thinking that you are moving forward. However, since you are in fact as static as the figure, you misperceive the two middle vertical lines as curved ones. Again, the time window again is as short as 80 milliseconds, but it is still enough to reflect life-threatening situations.

It is likely that we bank on this evolutionary adaptation of "future prediction" in the society we built. Moving faster than we can run when piloting a motorized engine (car, motorcycle, plane, be it for real or in video games where anything goes), as we navigate and avoid collisions. In sports, reaction times as fast as 300 milliseconds have been observed for table tennis (hitting a ball coming at 200 mph) or for soccer (goal keeper's reacting to a penalty struck at 80 miles per hour from the penalty spot). Auditory stimuli have been associated with even faster reaction times. Hundred-meter sprinter Bruny Surin, in the 1999 World Championships' 100-meter semi-finals, homologated a record reaction time of 101 milliseconds.

One can reflect on why the brain takes more time to generates visual realities compared to auditory ones. For a serious headache, how might the brain bind together sensory information with different processing time? The process is known as multisensory integration, and it is an exquisite feat involving thalamo-cortical connections (for conscious responses) as well as the deep layers of the superior colliculus (for unconscious responses). But enough digression; this is just to remind you of how so very tremendously precious your brain is.

Back to the two circles illusion (figure 1): how long did it take you to see it? There is seeing and *seeing*... How about the actual experience of being fooled into thinking that the two circles are distorted? We have established that Gabor patches of distinct orientations and positions are central players. And we said that your brain builds images (such as different faces) by linking an array of individual Gabor patches or small lines, each with their respective position and orientation.

David Hubel and Torsten Wiesel established that as visual information first arrives from the dLGN to the primary visual cortex (also called V1, striate cortex, or area 17), it is computed as line recognition. They inserted electrodes in the primary visual cortex of anaesthetized cats and presented visual stimuli through a glass slide projector on a screen placed in front of the cat. Instead of responding to round spots of specific sizes (centre surround receptive fields), they found that cells responded to lines (edges) or specific orientations.

David Hubel recounted their frustration of not being able to get cells in layers 2-3 of the primary visual cortex to respond to the conventional centre surround visual stimuli established in the retina and the dLGN. Instead, these primary visual cortex cells responded at unexpected instances, namely during the swapping of glass slides. They eventually figured out that electrical activity from these cells was in fact modulated by the projection of lines drifting at specific orientations: in their case, the moving edges of slides being removed from or inserted into the projector.

Researchers now tend to rely on sine-wave gratings (such as Gabor patches) to optimally replace these lines. Hubel and Wiesel received the 1981 Nobel Prize in Medicine and Physiology for their work on visual processing. A major achievement was their characterization of four principles of the primary visual cortex organization:

1) Retinotopic maps, that is, the relative position of retinal ganglion cells with each other over the retina surface, are reconstructed in the primary visual cortex; a notable feat considering that the retina map is lost as the 1 million RGC axons exit the retina, forming the optic nerve in humans. Of note here is that as in Penfield's homonculus, a proportionally larger area of V1 (more than half of it) is devoted to the fovea, which only represents 0.2% of its total surface (Polyak, 1941).

2) Ocular dominance columns; dLGN relay neurons carry segregated information from our respective right and left eye, forming columns as they arrive in layer 4 of V1 (information from respective columns is compared downstream for disparity), you guessed it; this is the basis of how we see in three dimensions.

3) Blobs for colour information processing; koniocellular dLGN relay neurons project axons directly to the blobs in layer 2-3.

4) Orientation columns: for each retina location, cells with the same orientation preference are regrouped into a column. Hubel and Wiesel proceeded to insert electrodes horizontally, elegantly showing gradual changes in orientation preference as the electrode

progressed along the horizontal tract, moving across columns.

Processing distinct properties of the visual image in parallel favours speed, but also optimizes the number of potential permutations, alike having separate ingredients to make a recipe as opposed to pre-mixed ingredients such as spices, which would limit the number of possibilities. What you see is a bit like the outcome of unique custom-made recipes.

Hammering the message (again unsubtly), that our brain generates what we experience as reality, and that illusions provide insights into the mechanisms producing what we know as reality: let us consider yet another illusion: the rotating snake illusion (figure 19).

Nothing is actually moving, yet the area of your brain responsible for detecting movements (V5 or MT) is active, generating your perception. Studying movement sensitivity in owls, Semir Zeki identified a specific region, which he named MT (equivalent to V5 in primates). Using functional brain imaging, Kuriki and colleagues (2008) showed that this region was activated when human subjects experienced the moving snake illusion. If you are using a computer to read this book, you can print the illusion on a paper and ascertain that nothing is moving. Which pathway is involved here? The magnocellular. Let us try the following experiment, paying attention to what happens when we change your fixation to different locations. The illusion of motion ceases precisely at the spot you are fixating. Why? Because fixating means that the image falls on your fovea, and thus you are now relying on the parvocellular pathway (fine details vision). The illusion of movement takes place peripherally to your fovea (outside your fixation area), where movement detection is optimal.

Figure 19

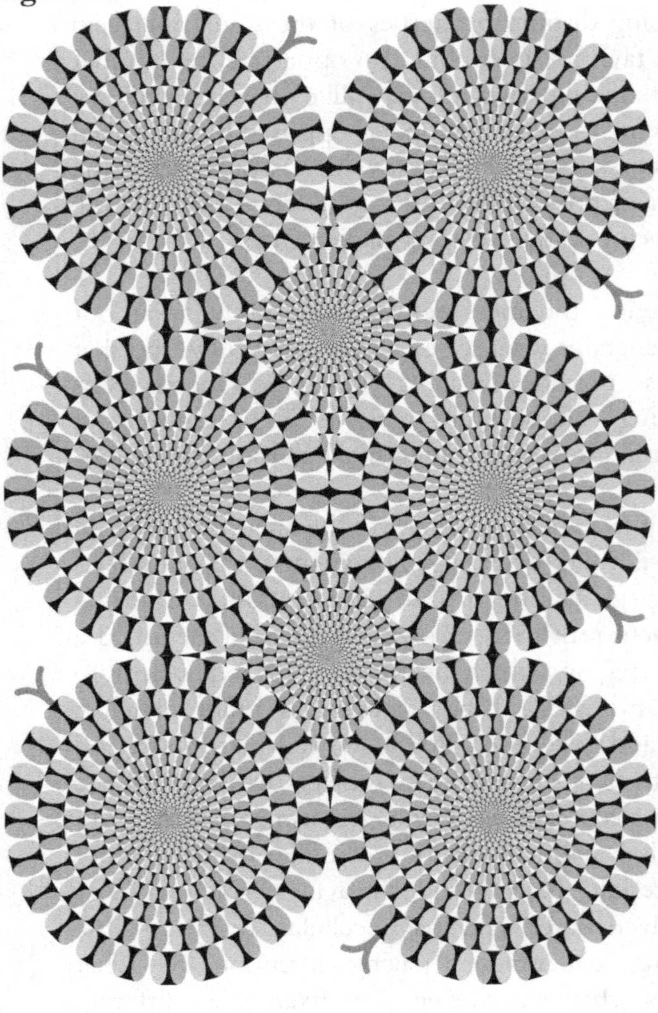

People with established lesions in cortical region V5 (equivalent to MT) have various levels of inability to perceive motion, a condition known as motion blindness or akinetopsia (Vaina et al., 2001). Experimental inactivation of V5 with transcranial magnetic stimulation (TMS) induces temporary akinetopsia in humans (Beckers and Zeki 1995). TMS is one more tool us humans have developed to guide us in our quest of understanding how our brain constructs what we experience as reality. Used in combination with functional assessment tools such as the EEG (electroencephalogram) and fMRI, TMS has provided novel relevant information on how individual neural networks interconnect (and even remodel, as in undergoing plasticity) to generate our thought processes (Rossini et al., 2019). Can you think of other approaches to glean insights?

Examples include: 1) studying stroke patients with impairments in highly defined brain areas; or 2) using drugs such as the hallucinogens contained in magic mushrooms (psilocybin and psilocin). By acting on a combination of serotonin receptor subtypes, psilocybin has been shown to induce motion in inanimate object without the use of illusions like the rotating snake illusion (Carter et al., 2004). In animal models, one can insert (using viral vectors) a gene coding for light-absorbing proteins that are related to neuronal excitation or inhibition, inside specific neurons. When illuminated at the correct wavelength, neuronal activity can be turned on or off. This approach is called optogenetics. In this manner, one can study the role of specific neuronal population on behaviours such as visually guided tasks in mice (Carrillo-Reid et al, 2019).

More illusions; no magic mushrooms required. We experienced with figure 19 the interplay between parvocellular and magnocellular, which is to say that looking directly at a pinwheel, stops it from moving. Now, I will to describe, rather than illustrate, a lesser-known illusion: the Cheshire cat illusion. I experienced it for the first time at the San Francisco Exploratorium (https://www.exploratorium.edu/snacks/cheshire-cat). You can go to their website for a detailed guide on how to make it yourself. In brief, you need to look at someone in front of you, directly with your left eye, while your right eye looks at a 45 degrees angled mirror that projects to a white wall on your right. You see the person in front of you using both eyes, but here comes a rivalry between the parvocellular (left eye) and the magnocellular (right eye) pathways. You create this rivalry by moving your right hand on the white wall. What happens is that your left eye sees your partner's face while your right eye sees your right hand moving on the wall. If seeing movement wins the rivalry (especially if you move your hand a bit in the periphery) then part of your partner's face will disappear (usually the eyes and nose, leaving a visible smile) solely when moving your right hand on the white wall.

Even more illusions; still no pharmacopeia. Try this: go to the toilet; look at yourself in a mirror, about 2 feet away, and move your eyes. With undeterred attention, observe your eyes as you move them, again and again, try again. Ask someone else to confirm that you are indeed moving your eyes. Take turns. Conclusion: you cannot see your own eyes moving. No matter how hard and briskly you move them, you cannot catch them moving. What just happened? You moved from one target to another, therefore, eliciting saccades, during which you were blind, yes blind. The proof? Well, you cannot see your eyes moving, doing saccades, because you are blind when your eyes undergo these saccades. We talked about the challenging statement that we are blind 4.5 hours every day (4 hours of saccades and 30 minutes of blinking); an icebreaker for a cocktail party. You can add the toilet trick, but concretizing it might require some creative social engineering on your part.

We see fine details by fixating with our fovea, an action we can call "foveating". Saccades are the jumps your eyes make between foveations, each covering 5 degrees of visual angle. During each foveation (between jumps), our eyes are in fact never still: they still move in microsaccades. Each microsaccade moves the centre of the fovea (foveola), which covers 1 degree of visual angle (Azzopardi and Cowey, 1993). There is no such thing as stillness when we look. Your brain generates an image based on a limited number of fixations (each consisting of 1 degree foveola sweeps with microsaccades) covering 5 degrees, providing the illusion that your visual world is evenly fine-detailed. In essence, your brain fills the gaps (jumps) between fixations.

The truth is, we are oblivious when it comes to how our brain generates what we perceive as being our visual world. Take the "Gorillas in our midst" illusion in which study subjects are asked to watch a video and count the total number of passes made between people playing with a basketball (Simon and Chabris, 1999). Prior to being shown the video, the study subjects are briefed on the very specific task ahead: counting the number of passes made by the team in white t-shirts. Subjects are prepared to accomplish their task astutely with utmost concentration, so much so that about half of the study-subjects fail to notice a woman in a gorilla suit walking among the group of people passing the ball between them. Once told, the disbelieving subjects are shown the video, now scratching their head: not only does the gorilla walk across the group, it stops in the middle to pound its chest proudly. Eye tracking studies showed that even fixation of the gorilla (occurring in 43% of the subjects) does not guarantee its awareness: half of these still (21% total) did not notice it (Gelderblom and Menge, 2018).

Inattentional blindness can present serious concerns in our society, such as in expert observational tasks including radiologists looking for anomalies in x-ray images. "*Even expert searchers, operating in their domain of expertise, are vulnerable to inattentional blindness*" (Drew et al., 2013). Not only are we unconsciously blind during the ubiquitous saccades and blinks, which we generate during our waking day (on average for a total of 4.5 hours) but we can also be blind to what we fixate, look at, and foveate. Heads up for when we will address what is consciousness: as you saw here, it is all very relative to the context in which you use the word *consciousness*; what is clear is that consciousness is not *something*, but rather a state. In fact it is many different states, again as you just saw with your eyes, above and to be continued below, using your fovea.

Our current society increasingly favours the fovea. The world we live in (and are changing with the technology we develop and implement) favours, depends upon even, foveal vision. Think of cell phones... the image falling on our retina occupies not even 0.2% of its surface. Social networking is more and more visual. We define our relation with others and ourselves by succinct, and often simplified, visual narratives. Image and video sharing platforms swarm: Facebook, Youtube, Twitter, TikTok, Instagram,

Pinterest, Twitch, Aparat, Bilibili, Vimeo, Mixer, Flickr, MySpace, Vimeo, Tencent, etc...

We rely on vision to understand reality. Our "view" of reality itself is biased. For instance, computer scientist Katie Bouman accomplished a feat previously considered impossible: she created an algorithm to build the first ever "photo" of a black hole. All matter and energy (including light, remember $E=MC^2$, where E is energy and C is the speed of light) can no longer be defined as such inside a black hole. Therefore, black holes are by definition invisible, hence their name. News media announced that the "photo" confirmed that black holes do indeed exist. What the image shows is actually not visible to the naked eye. Rather, the image is a graphical rendering of measures of the speed of matter (in a plasma state) around the centre of the black hole (M87), approximating the speed of light. In addition, the image has a diameter of 40 billion km and a corresponding weight 6.5 billion times the Sun's mass. At this point, the definition of matter, and even time itself, becomes irrelevant inside the black hole, where our ability to define reality, here with mathematics, falls apart. We simply did not evolve to interact with black holes.

If we are in awe with Northern Lights, it is purely accidental that we can see them; it just so happens that the magnetic storms on the sun generate electromagnetic waves on earth, which fall within the spectrum of human vision. Here is another, perhaps lesser-known, atmospheric phenomenon that ought to put you in awe as well: sprites or red sprites. These are large-scale electrical discharges occurring high above thunderstorm clouds, or cumulonimbus, giving rise to a quite varied range of visual shapes flickering in the night sky.

Vision (what we can see with our eyes) does not capture the complex nature of what we are able to observe, study, and are attempting to comprehend, black holes included. Mathematics alone, and their continued progress are a part of how we observe, study, and try to grasp reality, beyond the senses that we have evolved to possess. Some types of mathematics have no applications to the reality we currently fathom. The operative word here is "currently". One hundred years ago, Number Theory was considered useless. It turned out to be the foundational tool of computer network security: cryptography. Mathematics is not the first choice as a mass communication tool, at least not as much as visuals. Mathematics can speak through images, such as that of M87. Inspired teachers and communicators are precious individuals, much needed in our society for permitting us to progress beyond solely relying on what our senses have evolved to generate.

You construct the external world – you interacting with it – using limited physical aspects determined by how you have evolved from a single cell. Even within the evolutionarily determined limited range of physical modalities (such as so-called visible light), you see fine details only at a narrow angle of 5 to 7 degrees (foveation range). You are essentially blind to fine details beyond the horizontal 200 degrees and the vertical 135 degrees that your visual field spans (with both eyes open). To take in fine details over a larger area than covered by foveation, your eyes make saccades (during which you are completely blind).

Your depth perception (3D vision) is also limited by the overlapping of visual fields from both eyes, which cover 120 degrees horizontally. The most peripheral 40 degrees on each side are only seen by one eye, therefore rendering these areas of view devoid of depth perception.

Movement detection is also limited to a finite range; you cannot see the hour arm move on a clock. Forgetting all your racing thoughts, standing as still as possible, you can barely notice the minutes arm move. On the upper range, you cannot see the pales of a fan. Well, if you do not look at it directly and blink your eyes (to replenish your chromophore, *trans*-retinal, bound to opsins) you might convince yourself you had a glimpse of a pale; try it (you can do the same with the spokes of a tractor or truck wheel rotating slowly). The fastest processing time approximates 80 milliseconds for movement detection; however, other tasks, such as face recognition, entail an average of 300 milliseconds processing time prior to conscious experience: that is, before you "see".

Movies consist of 24 frames per second, equating to 1 frame presented every 0.042 second (42 milliseconds). Therefore, the duration of each image presented — 42 milliseconds — does not allow us "seeing" the images individually but rather as a continuous movement. Doubling the frequency of frame presentation as in the movie "The Hobbit: An Unexpected Journey", should in theory have no noticeable effect. Unexpectedly, informal reports by moviegoers relate perceptions of differences such as more vivid, at times distracting, details. Controlled studies (DoVale, 2017) confirmed the lack of consistent detectability in human perception at higher frequency frame rate (48 and 72 Hz) but showed "just noticeable differences" perceived by the study subjects as nosier signals. The take home message is that at 24 frames per second and beyond, consecutive frames become "fused", yielding the illusion of continuity.

When studying visual perception (the field of psychophysics), the term "fusion frequency" refers to the frequency of stimulus presentation at which it is no longer observed individually. Recordings of electrical activity from the retina (with the electroretinogram, analogous to the electrocardiogram for the heart) shows responses modulated by flashes of light presented up to 60 Hz (intervals of 17 milliseconds). Stop and think about the chromophore recovery cycle for cone opsins: a process involving over 20 different proteins and inter-cellular communication, which we will not explain here, but by all means feel free to search "retinoid cycle". The implication of cone opsin cycle recovery as fast as 60 Hz is that our retinas can process visual information much faster than it takes for the brain to construct a percept. In principle, neon lights oscillating at the frequency of the alternate current that activates them (50 or 60 Hz depending on the country) can modulate the electrical activity of your retina, but not your brain, at least not to the extent of leading to conscious perception.

Reflecting on fusion frequency, we might assume that the process of image generation, taking place in the brain, imposes limits to the information sent by the retina, but this is not always the case. Here is an example that defies intuition: hyper-acuity. Photoreceptor spacing limits acuity to 1/20 of a degree (achieved in the centre of the fovea, a region called the foveola). However, you can distinguish even finer angles (or spaces). Also called Vernier acuity test, the task is to determine whether you see one straight line or a "broken" line. The angle covered by this "break" is twice smaller than the limit dictated by photoreceptor spacing in the retina.

To understand how the brain achieves this feat (because this is a brain-based event), we rely again on Hubel and Wiesel's findings. Lines, not circles, are the language used by the primary visual cortex. Remember the edges of the glass slides that optimally modulate the activity of V1 neurons. The information about space discrimination sent by the retina (centre surround concentric receptive fields) is of lower acuity than that resolved by the brain itself. The brain is able to differentiate between continuous versus "broken" lines, which represents the foundation of visual information encoding in the primary visual cortex: V1 neurons respond (are "tuned") to lines of specific width and orientation; think of the Gabor patches.

Please proceed with care: more construction work ahead. Your brain has an obvious additional task, which you are aware of, the blind spot. I mean, you are not aware of the blind spot; this is the whole point of the task ahead. The retina is inverted, that is, photoreceptors (transducers of light into electrical changes) are at the back of the retina. They convey information to retinal ganglion cells (the only output to the brain), located on the surface of the retina. The axons of the retinal ganglion cells pass through the retina, all one million of them, piercing the retina (in a figure of speech) to form the optic nerve. The exit point is known at the optic disc, or optic nerve head:

this is the blind spot. Close your left eye, look at this square □; keep looking at it while trying to observe the tip of your right index as you move it slowly horizontally to the right, starting from the square. You will find a specific distance at which the tip of your finger disappears. This is your blind spot; it is real. Its location is about 12 - 15 degrees to the right of your right eye's visual field, and just slightly below by 1.5 degrees, to be precise. It covers a range of 7.5 degrees vertically and 5.5 horizontally. You must make deliberate efforts to experience your blind spot. Your brain "fills the gaps".

Not seeing, as in the blind spot experiment, does not imply experiencing blackness or being in the dark. No, you just experienced it: nothingness. The tip of your finger simply *disappeared*. This is what it feels like to not see at all.

By age 75, about 1 in 4 people will undergo progressive central vision blindness (loss of foveal cones) due to AMD (age-related macular degeneration). Perhaps lesser-known or under-reported are the hallucinations associated with the "release" of modulation of the visual cortical regions. As we approach Charles Bonnet Syndrome (also referred to as visual release hallucinations) with constructive curiosity, we can further our comprehension of how the brain generates what we see. Here is also an opportunity to develop kindness towards others and yourself when experiencing these initially quite disconcerting hallucinations. Visual hallucinations do not necessarily (in fact rarely), mean that you are loosing your mind. No, your mind (well your brain), is an active construction site.

Even more construction ahead. You see colour using three types of cones: red, green and blue. The koniocellular pathway represents already mixed colour information from these three types of cones interacting with each other in the retina. They converge information onto blue-yellow bipolar cells, which in turn talk with K retinal ganglion cells. But did you know that there are no blue cones in the fovea? How can this be? You see fine details using your fovea. When looking at something, you foveate.

Here is a potential conundrum, at first glance at least. If there are only red and green cones in the fovea, why is it that red-green colour blindness (occurring in 7 to 8 per cent of males) does not impede fine detail vision? The inability to see distinct red and green colours, respectively, does not imply reduced visual acuity. Fine-detailed vision is mediated by midget retinal ganglion cells (also called P cells, for parvocellular) with centre surround receptive fields that convey information from red versus green cones (red centre - green surround or green centre - red surround). The information encoded by midget (foveal) ganglion cells is achromatic. Red and green cones are stimulated differently at either side of the edges (or Gabor patches), but this information has nothing to do with colour at this point. Think about converting or even printing a red and green picture in black and white: the contrast will not vanish, and edges will be clear. Colour information is segregated from and analysed in parallel to information about fine details, alike separate ingredients prior to being blended into a recipe. The first convergence of these separate pathways (time to mix some of the ingredients) occurs in layers 2-3 of cortical area V1 where information about fine details (parvocellular pathway) is sent to the blobs (koniocellular pathway).

You might still ask, why can I see blue colour with the finest details if there are no blue cones in my fovea? To this, I will venture in saying that microsaccades likely have a role to play, and will stop here because more light remains to be shed on this matter. Of interest and perhaps somehow related, in 2017, a new anatomical discovery regarding the fovea was announced. Presented as such, one pays attention: how can we still be discovering novel anatomical features in the human body? The foveal region being devoid of blue cones (known as the Maxwell spot centroid) was not the discovery itself. We knew about it, as mentioned above: the fovea is devoid of blue cones. The novelty was that this topographically defined region, with clear borders and geometrical shape, tends to be evenly rounder in both eyes of people affected with dyslexia, while eyes in the control group presented with asymmetry between eye sides (Le Floch and Ropars 2017).

Readers who have cultivated an interest for visual perception and/or have read Oliver Sacks (The Man Who Mistook His Wife for a Hat) will have heard about the specific inability to recognize faces. Some of you might even have learned and retained the exotic name characterizing this high-level ventral stream of visual agnosia: prosopagnosia. In the spirit of further populating the list of examples supporting the view that the brain creates what we see, let it be known that prosopagnosia is linked with a dysfunction confined to a region of the cortex in which activity is specifically modulated in response to looking at faces: the fusiform gyrus. The ingredients have been blended intricately to make unique recipes, unique for each face. Convergence is high and exclusive to each memorized face. The list of ingredients is long; they were harvested using saccades and microsaccades, with blending occurring at specific stages. Elementary ones, mentioned above, include lines of specific orientations to make up the contours of the face and its key features: eyes, nose and mouth and facial hair when applicable.

Which species are capable of facial recognition? One would first have to define what a face is in order to objectively account for the wide range of species and their respective ecosystems. If we replace the word "face" with "specific individuals", we can include cattle (Coulon et al., 2009), reef fish (Siebeck et al., 2010), paper wasps (Miller et al., 2020), just to name very few of them. If we take a human species-centred (anthropocentric) position and ask which species can recognize human faces, we can still copiously populate a list: dogs (Mongillo et al., 2017), sheep (Knolle et al., 2017), fish (Kohda et al., 2015), and birds (Loidolt et al., 2003). This being said, such a question has absolutely no pertinence from an evolutionary perspective; animals, even domesticated ones, have not evolved to recognize human faces, and answering this question warrants receiving an Ig Nobel price.

To be fair, however, there is relevance from these findings at the level of the recognition of friends versus foes, enemies versus potential mating partners. One can also learn from studying the brain activity involved in performing these tasks. Comparative biology is a formidable tool when it comes to deciphering how the brain builds specific aspects of mental models. There is also an opportunity to take a step away from our tendency to be anthropomorphic: to ask questions that are not influenced by how we as humans take in the world.

The situation described here, that of our physiology, is a solution that arose from billion of years of evolutionary trial and errors. The driving force for why these physiological mechanisms exist at this point in evolution is because their multiple iterations have been compatible with survival and reproduction of our ancestors. With the exception of species "seeing faces", why focusing on us, *Homo sapiens*? Evolution has led to an extensive variety of other solutions.

Still talking about vision – always involving the conversion of photons into electrical changes – the term "visible light" varies between species, from UV to infrared along the electromagnetic spectrum. Remarkably, over the 3.8 billion years of evolution, the chromophore has remained remarkably stable: retinal (the most common), 3-dehydroretinal (interchanging with retinal inside the same opsin), 3-hydroxyretinal (found in insects), and 4-hydroxyretinal (limited to a single species, the deep sea squid). In all species, the chromophore isomerizes from *cis* to *trans* when absorbing the energy of a photon. It is the pigment type (opsin), with which retinal establishes covalent bounds, that determines the absorption spectrum.

Opsins are proteins; their respective genes encode their amino-acid sequences. These sequences underwent several iterations across evolutionary trials and errors. Most mammals are blue and green dichromates, with the exception of a few primates like us, human who possessing 3 opsin types: red, green and blue. We are trichromates. Even bulls do not see the colour red of the infamous matador's cape; they charge the cape not because it is red but rather based on perceived threats from the deft handler. On the other hand, some species possess many more cone types. The winners for the most number of types are the stomatopod crustaceans. For instance, mantis shrimps contain 12 functionally distinct cone types (Thoen et al., 2014). Despite their varieties, all opsins are g-protein coupled receptors with seven transmembrane domains (think of a thread passing 7 times through a textile).

In all species, unicellular (prokaryotes and eukaryotes) and multicellular (metazoans), response to light (photoisomerization) results in a shape change of the receptor (opsin), allowing binding of its g-protein and kick-starting the transduction pathway. Another modality associated with light is the polarity of photons: that is, the plane along which they vibrate as waveforms. When covalently bound to its opsin, the chromophore (retinal and its three close friends) is oriented parallel to the cell's membrane. With this particular configuration, light absorption is optimal when photons vibrate along the plane of the chromophore, which is the same plane as the membrane itself.

Because of this phenomenon, known as dichroism, the potential to detect polarized light is present in all species. However, only a few subsets of species, like bees, modify their behaviour as a function of light polarity. We cannot see polarized light, we cannot see infrared light, and we cannot see UV light. We can

- You: Why are there so many figures and details? So complicated. I thought this book was about the meaning of life... feeling it, whatever it might mean... are we there yet?
- Me: How do you feel?
- You: What?
- Me: How do you feel? How about your parents, their parents, your ultimate ancestors? Lets get back to single cells, 3.8 billion years ago.
- You: I am loosing my patience, what is the meaning of life?
- Me: Feeling is part of interacting with the environment around you... and inside you. But can you feel "inside your brain"? Can you feel the activity in your brain, the very activity that makes you feel? No you cannot. Neurons inside your brain do not have sensory receptors eliciting perceptions. In other words, you cannot feel their activity, so you cannot feel them; it is a mystery, as if cells in your brain had a life of their own, as if your thoughts had an origin beyond your body: as if body and mind were two things (duality).
- You: You are so very boring...
- Me: OK, here is what I am trying to accomplish here. The original meaning of the word "feeling" is: sensing, using our senses. We have evolved from single cells, which survived, exchanged genes, and reproduced, leading to you, yes you. The process has been continuous; nothing mysterious happened. DNA was passed on and evolved to the point that you can be here now: impatient, eager to have an answer to the meaning of life.

How so easy it is to get sucked into specific aspects; we must keep in mind the more global perspective of time and space (both interdependent), and other dimensions that are not necessarily linked with how we evolved. Let us take a step back, being mindful of the context of how we have evolved, and the reality we can fathom.

Chapter II

Evolution Of The Senses

"*Nothing in Biology Makes Any Sense Except in the Light of Evolution.*" Theodosius Dobzhansky

According to Hoffman and Singh (2012), evolution "*shapes perceptual systems to guide fitter behaviour, not to estimate truth*". They state that vision "*rather favours fitness of the organism with the external world*". Illusions support this assertion: we perceive reality not only based on how we have evolved, but also on how we have been raised culturally and socially. To truly learn is in fact to unlearn cultural and anthropomorphic biases so that we can observe without ego, false threats or expectations. There is a fundamental precept in evolutionary biology known as Morgan's Canon, which warns us against anthropomorphic biases when trying to understand the behaviour of animals. Basically, we should avoid giving intent to the process of evolution and its outcome, let alone subjectively interpret the process from our own human and culturally biased perspective.

We have a limited perception, comprehension, imagination, and perspective. If I say "trees are our distant (OK, very, very, extremely distant) relatives, with bacteria being our common ancestors" The statement is accurate; you are related to anything made of wood. In the so-called civilized world, we are not used to thinking in this manner. What if I were to say that evolution continues? Think of horizontal gene transfer where viruses (bacteriophages) mediate gene exchange between bacteria, underlying bacterial resistance to antibiotics. If I say that evolution does not necessarily mean going toward more and more complex forms, would it come as a surprise? Well, I just said it: evolution continues, and can occur in so-called simple species such as bacteria and viruses.

You might have heard, read, seen somewhere, or perhaps even have considered that humans are the most evolved of all species. We are inclined to think of ourselves as being the most complex life forms: the top of the top, the cream of the cream of the evolutionary tree. Look at what we can do: the world we build and destroy; evolution does not necessarily result in increased species complexity. A population can evolve to become simpler, such as having a smaller genome with fewer proteins and an overall more simplified physiology (Moran, 2002). On the other hand, possessing a higher number of genes does not necessarily predict what we might consider as complex. A tiny translucent water flea called *Daphnia pulex* holds the current record for the highest number of genes. I say "current", because this barely visible critter is the only crustacean species that has had its genome sequenced at time of writing. The results yielded close to 31,000 genes (Colbourne et al., 2011. Humans, in comparison, have about 19,000 genes (Ezkurdia et al., 2014). Not only does gene count fails to predict the evolutionary level, it does not reflect the genome size. The total number of genes is independent of the amount of non-coding sequence (genome size). The current record for the largest genome goes to the marbled lungfish (*Protopterus aethiopicus*) with 133 billion nucleotide base pairs

(Leitch, 2007), compared to 3 billion in humans (or 6 if you account for the chromosome doubling).

New forms of genes arise purely by chance. In addition to being passed down through multiple generations (vertical transfer), genes can be passed on between species (horizontal transfer, as for antibiotic resistance). Furthermore, to paint a more comprehensive picture of how species evolve, one must consider other events mediating horizontal gene transfer, such as bacterial conjugation, genetic symbiosis, hybridization, and chimera formation. Simply put, my goal here is to acknowledge that evolution is not merely the passing of genes from one ancestor to its descendent. That being said, we can assuredly assert that humans and all living species (including all of which has roamed the earth and is no longer among us) arose from single cells. Let us keep in mind the continuous aspect of evolution, albeit continuity equates a tortuous path with many roads diverging and converging, with links, bridges, tunnels, even airways and underwater ways, but enough metaphors. The genes, which underlie who we are, are an outcome of evolutionary events that began in single cells 3.8 billion years ago. Nothing magical happened; no external agents or intelligent design required here. Stop and reflect on 3.8 billion years. Take a deep breath or two or as many as you need to fathom 3.8 billion years. Did I write it clearly and simply? Here goes: 3.8 billion years.

The nature of the genetic baggage, of its preservation and its passing on — in sometimes tortuous ways — depends on encoding for novel proteins ultimately synergizing their actions, allowing the survival and reproduction of a species. To survive, species must sense their internal and external environments, and respond accordingly (modulating metabolic and sensorimotor pathways). It took 3.1 billion years of evolution for single cells to become dependent on each other for survival and reproduction. Grouped as one organism, cells became specialized. Sensorimotor pathways are comprised of distinct sensing cells and motor cells. Between these cells, interneurons (inter: between) arise to form primitive nervous systems (nerve nets). Assemblies of interneurons generate a representation of the environment in which an organism survives and reproduces.

Warning: A hurdle to our grasping of reality is that we are self-centred; we are the hubs: the reference point. We are anthropocentric, thinking of ourselves as separate from and superior to nature. You may have read something alike "the human brain is the most complex entity we know of." With 90 billion neurons, up to fifty times more glial cells, and the number of synapses estimated at the number 10 followed by 14 zeros, the brain certainly puzzles us and drives us. It drives our curiosity, our desire to explore its function, and our potential as a society. Perhaps one reason we obstinately consider the brain as the most complex entity is that we think about how we think, about how the brain does it. While the brain might be special to us humans, one can find complexity and be in awe with so many other aspects of the world we discover. Thinking of thinkers who taught us to think otherwise, three names come to mind. Copernicus, and how the earth is not the centre of the universe; Darwin, telling us that we share common ancestors with chimpanzees; and Freud, who states that we are oblivious to our thinking processes, since these are very much a part of our body. We treat "consciousness" as a feat, something unique, something mysterious, and something beyond reductionism. Consciousness did not "arise" at a fixed point during evolution. Consciousness is a continuous process of evolution. Evolution is a consequence of DNA mutations, which led to unicellular and multicellular species; it has no purpose, no agenda. Evolution has no intent. We certainly are not a goal in itself, and we certainly are not central to the process of evolution. We could say we are passers-by, or even that we are an incident of evolution; we are incidental. Species are only conducts and vessels allowing DNA to mutate. As long as there is survival and passing of DNA, DNA will change and lead to other types of conducts, vessels, and passers-by. DNA itself has no purposes; it just changes with evolving hosts.

A short history of the senses

Figure 20

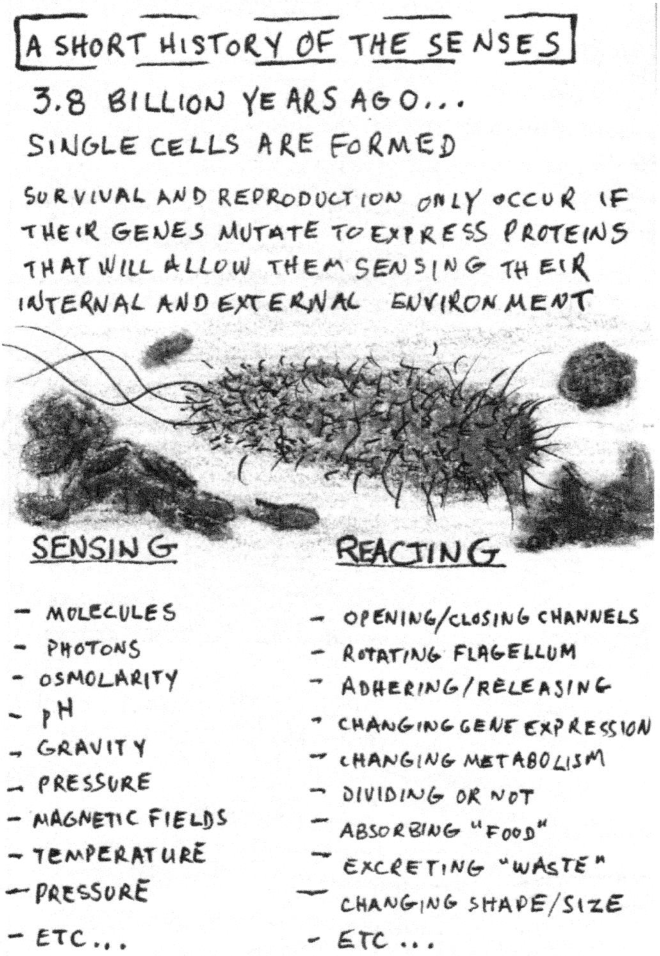

At its root, the term "to sense" signifies to transduce, to convert one form of energy to another form. Transduction dictates how species (unicellular, multicellular, colonies) respond to internal and external stimuli for ensuring survival and ultimately passing DNA, as it has occurred over the past 3.8 billion years (I said it again), and is still occurring.

Transduction will be our key protagonist in the story of how we came to be. We will focus on *sensing* for this short history of the senses, principally as a means to recount the story us. Us, the self-centred species, asking so many questions: why does the world exist, why are we here, who are we, where do we come from, and yes, what is the meaning of life? Hint: we do so because we can; our inclination and ability to do so are incidental to the evolution of predictive mental models, which emulate and predict what we experience as reality in a highly dynamic fashion.

The short history of the senses spans 3.8 billion years: a period that allowed a generous amount of time (as we are trying to conceptualize by breathing slowly) for trial and error. Our hero, transduction, underwent mind-boggling iterations. However, a single exception stands among all aspects of transduction: retinal. Retinal, the most common of 5 closely related types, has been and remains the chromophore used by all species, while opsin proteins epitomize the rich array of diversity of varieties.

You will see again; we already talked about phototransduction at length; we saw that when absorbing a photon, retinal isomerizes from *cis* to all-*trans*, inducing a shape shift in the opsin protein, to which it is covalently bound; we also saw that shape shifts allow G proteins (transducin is one many in this also large family) to bind their opsin receptor, kick-starting phototransduction; what we did not see was that in our ancestors, opsins (bacteriorhodopsin, with over 800 known types) relied on retinal to pump protons. Therefore, throughout evolution, retinal has played a role in sensing light and pumping protons (positive hydrogen ions, H^+). What does "pumping protons" do? To be disclosed in the first act.

Act I, scene 1:

We are at 3.8 billion year in the past. Our ultimate ancestors, prokaryotes (pro = before, karyotye = kernel or nut, before nuclei), live in undersea hydrothermal vents, shamelessly spurting copious amounts of sodium and protons (Martin and Russel 2003), which provides a limitless source of energy. Deprived of oxygen (photosynthesis has not evolved yet), the atmosphere lacks an ozone layer, allowing high-energy ultra-violet (UV) rays to bombard the earth and radiate under the great primordial brine. The trick of producing energy is accomplished thanks to retinal, bound with a bacteriorhodopsin. The union is known as the Mrp (multiple resistance and pH) antiporter, which transduces light energy into electrical changes.

In the cradle of life, our main protagonist, transduction, plays a crucial role. The transduction of light into electrical changes produces energy usable by cells. The shape shift ensuing from photoisomerization allows one sodium ion (Na^+) to get out, and for one proton (H^+) to enter the cell, an event referred to as "antiporting" or "proton pumping" (Lane and Martin 2012). While pumping protons itself does not produce energy, a gradient accretes along which electrons can flow. The flow of electrons through electron transport systems is used to store energy in its most ubiquitous currency form: ATP.

At this point in evolution, phototransduction is linked with energy production, rather than with "vision". At the same time, proton pumping leads to the formation ancient proteins: ATP synthetase, an enzyme involved in storing energy as ATP, and ferredoxin, a protein capable of metabolizing another compound also generously squirted by the undersea vents: iron (Lane and Martin 2012). Iron plays a key role in oxidation-reduction processes; it is an integral part of key proteins that constitute electron transport systems.

Act I, scene 2:

We are at 3.4 billion years in the past. It will take an additional 1.4 billion years of evolution for our protagonist, transduction, to mediate sensing. Imagine what can happen in the mean time. Feel free to take more deep and slow breaths to grasp the scale of time: 1.4 billion years. Prokaryotes are still the only life form of earth. They live and survive as colonies, secreting a type of glue (extracellular matrix), forming multi-layered biofilms: these few-centimetre-thick microbial mats that retain nutrients. These colonies release minerals, forming sedimentary structures known as stromatolites.

Bacteria and archaea cover the surface of the earth, exchanging information (RNA and DNA) between them using horizontal gene transfer in three different ways. The first way was given a name that might be confusing, as it is the same as our protagonist, "transduction". However, it is all about context; when referring to the injection of foreign DNA by a bacteriophage (a virus specific to bacteria) into the host bacterium, the term "transduction" should be used.

The second way is called conjugation; in this context, forget about verbs, but think about the transfer of genetic material between two bacterial cells in direct contact with each other.

The third way is "transformation", where the genetic material passes through the extracellular environment and is taken up by a bacterium (the singular of bacteria, in Latin). Together with spontaneous mutations of genes carried by viruses and resulting from UVB exposure (a powerful mutagen), these various ways of exchanging genetic information horizontally are responsible for the evolution of numerous novel genes.

It will take this whole timespan, 1.4 billion years, for photosynthesis to evolve. Bacteria transduce near-infrared light to produce sulphur or sulphate compounds rather than oxygen; their pigments (possibly bacteriochlorophylls) are ancestors of chlorophyll. Other non-opsin photopigments will also evolve. Photolyases and their descendants, cryptochromes, will transduce blue and UVA light to repair RNA and DNA damages caused by exposure to UVB light. These proteins use flavins (such as the vitamin B2, known as riboflavin) as chromophores instead of retinal. Cryptochromes (from the Greek κρυπτός χρώμα, "hidden colour") also regulate the metabolism of cells as a function of the daily variations in exposure to UV and blue light, a physiological process known as photoperiodicity or circadian rhythms, which has been conserved through evolution.

The ability for our ancestral cells to generate electrochemical gradient, and the ensuing energy production and storage (ATP), means they can survive outside of vents and independently of colonies. There were two distinct migrations, each leading to the respective and distinct evolution of bacteria and archaea. Bacteria and archaea evolve separately and outside of hydrothermal vents. Bacteria and archaea start evolving genes that promote motility, using specialized motorized propellers (fuelled by redox reactions) called flagella and archaella, respectively (Miyata et al., 2020). As we will see, alongside motility, diverse senses evolve; we are almost there. Just one more paragraph to breathe some fresh air, and we will see.

Act I, scene 3:

By about 2.7 billion years ago, bacteria and archaea have evolved from our Last Universal Common Ancestor, affectionately called LUCA. Cyanobacteria absorb visible light with a mix of pigments (phycobilins, carotenoids and several forms of chlorophyll). These pigments allow for the first time ever to use water (H2O) instead of hydrogen sulphide (H2S), or other compounds, as a source of electrons and hydrogen for fixing carbon dioxide (CO2). This revolutionary mode of producing energy and organic compounds (using only light, water and CO2) will change the earth forever. Why? The waste product of this process is oxygen. From 2.7 to 2.4 billion years, oxygen accumulates in the earth's atmosphere. The outcome of using light in this manner has been elemental in determining the course of evolution.

Act II, scene 1:

We are now at 2.4 billion years in the past, when sensing likely occurred. The reason we think "likely," is because genetic studies (genomics) point to the first appearance of a gene coding for halorhodopsin at that time (Armitage et al., 2003). Halorhodopsin, another ancient type of opsin, found in halobacteria, is a prime candidate as the first "seeing" device. A confusing name, since halobacteria are not bacteria but rather a type of archaea (from the Greek ἀρχαῖα, arkhaîa, "ancient"). Halorhodopsin possesses a role in both energy production as a proton pump and in light-detection as a sensory transducer. By combining both functions, halorhodopsin is ideally placed to allow phototaxis (photo = light, taxis = movement), that is, light-guided (sensory) movement (requiring energy). Halobacteria might very well have been the first species capable of "vision". The term vision here implies detecting light — intensity, polarity and wavelength (colour) — and moving toward or away from its source, along its gradient. Such behaviour would have determined life or death in these single cells.

There is also evidence that halobacteria are able to "smell": that is, they recognize specific chemicals (serine) and can alter their swimming behaviour accordingly (chemotaxis); see figure 21. How do we know this? Fortunately, species similar to our first ancestors have been able to survive. Halobacteria and many more archaea species flourish today in most habitats, including soil, oceans, and marshlands, and are even a part of the microbiota (species living in our guts). For instance, in humans, archaea are found in the gut, mouth, and on the skin. Now add to your grasping of the richness of life forms the fact that species from the prokaryotic kingdom (bacteria and archaea) found in present times represent only a minuscule portion of all prokaryotic species thought to have existed on earth. While today's prokaryotes have evolved, their study (such as genomics) does shed light on how sensing was achieved in our ancestral single cells.

Act II, scene 2:

We are still at 2.4 billion years ago, and in addition to photoreception (opsin bound retinal), prokaryotes also have evolved to sense diverse aspects of their internal and external environment, and to react accordingly. We can state that this marks when "feeling", in terms of "sensing" the environment, truly evolved. Here is a tentative categorization of what single cells sensed (transduced): 1) mechanical changes, such as vibrations, pressure changes, and contacts with structures including crowding and quorum sensing (biofilm formation); 2) molecules (chemoreception); 3) osmolarity (concentration of molecules); 4) acidity (pH); 5) magnetic fields; 6) gravity; and 7) temperature. To the list, let us not forget number 8: photons. These sensory transductions were associated with various responses, or behavioural changes by the cell. Here is another list, now for the ways single cells could respond ("effectors") to various stimuli: 1) opening/closing channels; 2) rotating flagellum to move away from threats or towards food; 3) adhering or releasing from a substrate; 4) changing gene expression; 5) changing metabolism by being active or dormant; 6) dividing or not; 7) excreting waste; and 8) changing shape/size, etc.

So, we have a list of sensors and a list of effectors. Many permutations are possible between the type of sensors (8) and the type of effectors (8). In the likely incomplete list above, the theoretical tally comes to 64 possible sensing (input)-response (output) types. Instead of considering all possibilities, let us examine how sensing determines behaviour in prokaryotes.

There is still so much more to talk about what happens at around 2.4 billion years ago. This is the sensing boom, the explosion of sensing. If we stop to think about it, if we are following the lead of how our ancestors sensed and felt, then this is the time to stop and examine more deeply.

Let us time-travel to 2.4 billion years ago; how do single cells devoid of any nucleus (no nuclei in sight until 1.85 billion years ago) feel? We have lists of what they can feel and do, but how do they accomplish it? This is the key question, if indeed at some point we will address what is the meaning of life. We would like to know how "we feel" and how it all started, because nothing supernatural happened from the first instance of "feeling" to the evolution of the self-centred us. Evolution is a continuous process. We know not much compared to multicellular species. The word explosion is in fact used for evolution, not here at 2.4 billion years ago, but rather much later at 0.541 billion years ago when began the Cambrian explosion, during which most animals evolved. It is typically during this period that focus is put on the evolution of the senses.

There are copious fossil records to study how animals might have sensed, and modified their behaviour accordingly. Note the time 0.541 compared to 2.4 billion years. What do you notice? The level of precision is different. Even the use of 0.541 is typically considered as approximate in scientific circles. Well, everything is relative; if we were to use the same lens of scrutiny for the first evolution of the senses, we would be disappointed by one decimal digit "point four" of 2.4 billions. We should, point out that the first decimal (number four) is in the order of 100 million years; this is the level of precision. We fall into the trap of convenience, thinking that during this time, precision of time and related chronological sequences do not matter as much. We are not only human-centred, but also time-centred: we are chronocentric. Humans started evolving at the end of a 13.8 billion-year process for the cosmos to emerge out of the big bang. As proposed by Carl Sagan, if we were to consider a one-year calendar for the history of the universe, then, recorded history would represent the very last 10 seconds of the cosmos' existence. Einstein also told us that time was relative.

We do not know the exact time at which senses evolved. The number 2.4 billion is very much an approximation. Now, pay attention, because we need to set our time machine to interview our first single cell, a bacterium, so we can ask how it feels. If you think bacteria do not speak, think again (Bassler and Losick 2006). What do you suggest for time setting? How about 2,416,722,893 years ago, a more precise number than 2.4 billion, picked just randomly. Which month, day, and time? How about 22nd February, 14:00. As for the location, it will have to be underwater, so I how about the Nuvvuagittuq belt in Québec, Canada, which is the location of one of the oldest organic sediments that arose from life in thermal vents (Dodd et al., 2017)? Surely we'll meet an archaeon or a bacterium, which has quite a story to tell, and a very long one at that.

Let us embark in our time machine. But one important notice: we must disinfect our devices and gown up to prevent contaminating the ancestral environment, as we individually carry over 3.8 trillion bacteria (across over 10 thousand different species), and even more viruses: 380 trillion to be almost precise. Such uncommonly out-of-range levels necessitate special units of measure; in French it is "shitload", as isn't this after all where most bacteria end up: in faecal matter (think microbiome)? Did I also mention that our precious and wonderful human body also hosts archaea, protists and fungi? We are ready to leave, but if you ever have second thoughts about this trip, consider that perhaps the answer of how prokaryotes sense is right here in ourselves: in our very own faecal matter. As we examine the nature of transduction in prokaryotes, we will obtain evidence for evolutionary conserved mechanisms. But first we must leave: past time looms ahead.

Act II, scene 3:

It is the 22nd of February, 2,416,722,893 years ago, 14:00 (or 2PM if you prefer). We are underwater, in the Nuvvuagittuq belt, Québec, Canada. Here it is: a typical prokaryote. Immediately noticeable are its numerous cilia and a flagellum (figure 20). Really, immediately noticeable? Well, yes: figure 20 is obvious, the drawing does instruct. As is the case for all bacteria (and archaea) living today, the problem here is that we cannot observe them visually, at least not directly with the naked eye.

We are visual when it comes to grasping our environment. Remember the black hole. What we need are advanced imaging techniques, such as electron microscopy and cryo-electron tomography. Both are invasive, requiring the killing and dehydrating of our subject (Watts et al., 2019). We could go for super resolution spinning disc confocal microscopy, but again we would have to interfere with the cell, and the resolution would not allow "seeing" inside the cell.

I mentioned an interview so as to lure you here in our time-travel capsule, because we are also auditory. The bacteria won't speak directly; however, scientists have injected specific genes in bacteria (transfection), which code for gas-filled protein nanostructures, which can be visualized using ultrasounds (Bourdeau et al., 2018). So, we could track these transfected bacteria using echolocation, bouncing ultrasound waves through cells. The resolution is again poor; allowing for localizing bacteria as colonies, but not single ones, let alone not intracellularly. Therefore, we will learn how bacteria "feel" by combining numerous approaches, but not by looking at them directly; we will need to interfere with them and likely killing them ultimately.

To continue our exercise, what do we expect from the information scientists have gleaned so far using various complementary approaches (imaging, molecular biology, genetics, biochemistry, computer modelling, bioengineering, nanotechnologies, etc.)?

The bacterium we are meeting is a *Halobacterium salinarum*; the holder of what might have been the first sensing device, halorhodopsin, but also the holder of other opsins such as bacteriorhodopsin and sensory rhodopsins (I and II); an ideal species since it has survived all these years, and we can still study it today.

The Red Sea owes its name and colour to the presence of bacterioruberin, a bright red pigment found in the membrane of *Halobacterium salinarum.* That is not all, pink flamingos also owe their name and colour to *Halobacterium salinarum*, which they consume indirectly by filtering brine shrimps, who themselves feed on these archaea. We could pass on warm greetings from our study subject's descendants, but genealogy in the bacterial world is a nightmare to track, essentially because DNA is exchanged horizontally, using viruses and the engulfment of DNA containing mitochondria, as well as other effective methods. *Halobacterium salinarum* is particularly gifted at modifying its gene sequences by using insertions and deletions. *Halobacterium salinarum* has persevered as a species over 2.4 billion years.

Let us first consider phototaxis and chemotaxis (figure 21). For a more detailed schematic please consult Suzuki et al. (2010). Disclosure: while the events described here apply to *Halobacterium salinarum*, some of them rely on faecal matter as a source of inspiration (*E Coli*). Both the visual and chemical stimuli utilize the same phosphorylation cascades in the cytoplasm, and eventually, a phosphorylated response regulator, CheY-P, binds to the C-ring of the archellum (archaean equivalent of bacterial flagellum), which controls the direction of the rotation.

Figure 21

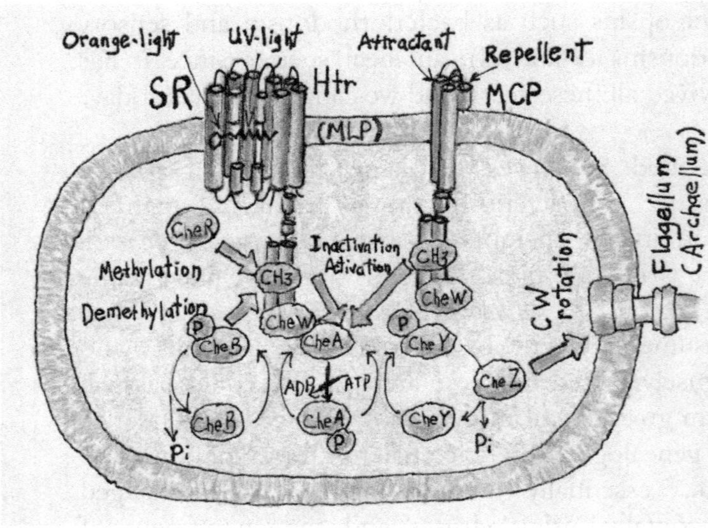

Light stimulation activates sensory rhodopsins (SRs I and II) and isomerizes retinal (the chromophore) from *cis* into all-*trans* (as illustrated for human in figures 8, 9, 10, and 11). Like all opsins, SRs are G protein-coupled receptors with 7 transmembrane domains that covalently bind retinal as their chromophore. Once shape-shifted, SRs can now bind their G-protein, the halobacterial transducers Htrs, which form a complex with CheA and CheW. This complex can then activate phosphorylation cascades that modulate the autokinase protein CheY and control the direction of rotation of the archellar motor. Activation by UV-light positively regulates CheA, resulting in negative phototaxis (CW, clockwise archellar turn, swimming away from the light source), protecting the cell from the deleterious effects of

UVB light. Stimuli in the red-orange spectrum lead to the positive phototaxis (CCW, counter clockwise archellar turn). The resulting movement attracts the archaeon to nutrient sources and brings it closer to light sources. Light can be transduced into energy by pumping chloride ions into the cell. In this manner, cells can live under anaerobic conditions (without oxygen). For chemoreception, the cognate (similar) chemicals, such as serine (attractant), bind to the extracellular domain of the chemoreceptors (MCP), and the binding induces the structural changes of MCP. The signal transfer pathway of MCP is thought to be similar to that of the phototaxis. The binding of chemicals that are toxic tends to lead to negative chemotaxis (clockwise archellar turn), while binding to nutrients such as serine typically leads to positive chemotaxis (CCW, counter clockwise archellar turn).

Act II, scene 4:

In addition to phototaxis and chemotaxis, *Halobacterium salinarum* can also sense differences in membrane potential by using the transducer mpcT, and can sense air bubbles (aerotaxis) by using the transducers htrVIII and hemAT. Instead of going into details with regard to these mechanisms, let us take a broader view at all the types of sensing we expect to have evolved 2.4 billion years ago. Our ancestors already displayed all known modes of transduction existing today: 1) one-component systems; 2) two-component systems; 3) multi-component systems; and 4) Extracytoplasmic Factors (sigmas and anti-sigmas).

1) One-component systems are single proteins possessing both sensory domains (input) and regulatory domains (output). These proteins are for the most part located inside the cell (cytosolic, free floating). They recognize an input with their sensory domain (one or more) and transform it into an output with their regulatory domain (also could be more than one). Based on genomic data from thousands of species of bacteria and archaea, Ulrich et al. (2005) determined that this simplest form of transduction represents the majority of sensing events in prokaryotes. The main role of a one-component system is to maintain homeostasis (overall cell health) by binding metabolites with its sensing domain, and by binding DNA (readily accessible since there is no nucleus) in its regulatory domain (DNA-binding helix- turn-helix, HTH, domain) in order to control protein synthesis.

There are over 20 distinct families of one-component systems, with likely many more to be discovered. On-going investigations have yielded several hundreds of different types of proteins within a single family, each possessing distinctive properties and roles. Here is a list of the main families with their respective regulatory role (obtained from Cuthbertson and Nodwell, 2013):

AraC/XlyS: pathways for the catabolism of various sugars
ArgR: amino acid metabolism
ArsR/SmtB: metal homeostasis
AsnC/Lr: amino acid metabolism
Crp/Fnr: many cellular processes
DeoR: sugar metabolism
DtxR: metal homeostasis
Fur: metal homeostasis
GntR: numerous cellular processes

IclR: carbon metabolism
LacI: carbon metabolism
LuxR: quorum sensing
LysR: many cellular processes
MarR: antibiotic resistance
MerR: metal homeostasis
MetJ: many cellular processes
ModE: metal homeostasis
PadR: poorly characterized
TetR: antibiotic resistance
Xre: various cellular processes

Cashin et al. (2006) identified over four thousand different bacterial one-component proteins involved in the utilization of L-arabinose (AraC) and lactose (LacI), respectively, as carbon and energy sources. This means that there are four thousand different proteins possessing input and output domains dedicated to the utilization of two sugars!

More, more, more, you'd like more lists? Just one more list, again with the goal of providing a humbling experience of the complexity and diversity of one-component systems: the LysR family, which is described on the list above as being involved with many cellular processes. Maddocks and Oyston (2008) provided just a glimpse of the principal regulators (output) in this family, alongside the respective function of the gene they regulate:

AlsR: Acetoin synthesis
AmpR: beta-Lactamase
ArgP: Arginine transport
BenM: Aromatic compound degradation
BlaA: beta-Lactamase
CatM: Catechol catabolism

CatR; Catechol catabolism
CbbR: Carbon dioxide fixing
CfxR: Carbon dioxide fixing
ChiR: Chitin binding/chitinase
CidR: cid operon (murein hydrolase regulation)
ClcR: Chlorocatechol catabolism
Crga: Pili/capsule synthesis
CynR: Cyanate detoxification
CysB: Cysteine biosynthesis
CysL: Sulphite reductase
GltC: Glutamate synthase
HupR: Haem uptake
HvrB: S-Adenosyl-L-homocysteine hydrolase expression
IlvR: Isoleucine/valine biosynthesis
IlvY: Isoleucine/valine biosynthesis
IrgB: Iron-regulated virulence factor
LeuO: Bacterial stringent response
LrhA: Flagella, motility and chemotaxis
LysR: Lysine biosynthesis
MdcR: Malonate catabolism
MetR: Methionine and cysteine transport/biosynthesis
MleR: Malolactic enzyme
MtaR: Methionine transport
MvfR: Pathogenicity regulator
Nagr: Naphthalene catabolism
NahR: Naphthalene/salicylate catabolism
NhaR: Na+/H+ antiporter
NocR: Nopaline catabolism
NodD: Nitrogen fixation/symbiosis
OccR: Octopine catabolism
OxyR: Oxidative stress response (peroxyde)
PhcA: Virulence regulator
QseA: Quorum sensing
RbcR: Carbon dioxide fixing

RovM: Invasion/motility, virulence
SpvR: Virulence factor synthesis
SyrM: Exopolysaccharide synthesis
TcbR: Chlorocatechol metabolism
ToxR: Quorum sensing
YofA: Cell division
YtxR: ADP-ribosyl-transferase toxin

I stop here; you get the idea. This is likely the tip of the iceberg when it comes to the repertoire of one-component systems and their characterization. Like most types of scientific inquiries, bacteria and archaea signal transduction research is essentially funded for health related targeted goals, and rarely for the sake of curiosity and knowledge alone. In societies where even healthcare becomes privatized, how can one expect basic, as opposed to applied science (for targeted goals), to be valued? The irony is that applications often derive from non-targeted basic science; think of radar, microwave and GPS, just to name a few.

"*The scientist does not study nature because it is useful to do so. He studies it because he takes pleasure in it; and he takes pleasure in it because it is beautiful.*" Henri Poincaré

2) Two-component systems consist — you guessed it —of two distinct proteins that respectively handle the input (sensor) and the output (regulator). The former is a histidine kinase protein receptor (membrane-bound) for sensing, while the latter is an independent response regulator (cytosolic) that produces an output. Two-component systems are most common in bacteria (Gram-negative and cyanobacteria), where they allow them to sense, respond, and adapt to a wide range of environments, stressors, and growth conditions. In fact, histidine kinases and response regulators represent the largest gene families in bacteria. When stimulated, the sensory kinase phosphorylates itself. Of note, kinases are the name given to enzymes that add phosphate molecules (phosphoryl groups) to a protein, typically altering its function. Once phosphorylated, the sensor is activated and can now bind to a response regulator and transfer its phosphoryl groups at one or many locations. The addition or removal of one or many phosphoryl groups, their location, and their rate of intervention all contribute to controlling gene transcription (output). We can add one more aspect of signal transduction here with the two-component system as opposed to the one-component system, memory.

Yes, memory, in that very sense of remembering; the bacterium we are meeting, can remember. How? In 1997, Koshland proposed that the duration of phosphorylation (a higher-energy level) holds information about the state of the environment that is "*no less real or useful to the bacterium than the memory of humans is to their behaviour*"; quoted in an exhaustive review by Pamela Lyon (2015), entitled "The cognitive cell: bacterial behaviour reconsidered".

3) Multi-component systems are basically two-component systems that involve non-kinase receptors (such as methylated receptors in chemotaxis and phototaxis as we discussed above) to activate protein kinases, forming a relay of additional phosphorylation steps with other regulatory components and even allowing interactions with other signaling pathways. Serving its own purpose of phosphorylation as a key event in signal transduction our cell must find extracellular sources of phosphate. It is a bit like saying "I eat so I can eat". Our bacterium senses phosphate so that it can sense phosphate, but for many other reasons as well. What our cell senses is orthophosphate, an optimal inorganic source of phosphate. Are you sitting down comfortably? How many components does it take for our cell to sense orthophosphate? Seven. Seven distinct protein components: 1,2) PhoR histidine kinase and its response regulator PhoB (together forming a two component system); 3,4,5,6) four components of the ATP binding cassette controller (ABC) transporter Pst; and 7) PhoU, a chaperone-like protein that inhibits PhoR/PhoB (Hsieh and Wanner, 2010).

4) Extracytoplasmic Factors directly bind to an enzyme involved in protein synthesis from RNA: RNA polymerase. Their activity is controlled by anti-sigmas. Together, sigmas and anti-sigmas form a repository from which the cell can choose to optimally respond to different types of stresses, in a manner analogous to neural networks with on and off switches, noise level detectors, and signal modulation (from self-repression to amplification).

Act II, scene 5:

Just before concluding our time-travel at 2.4 billion years in the past, there is someone else I would like us to meet, a bacterium. We already spent considerable time with an archaeon, so why not meet a bacterium, not any bacterium, but one that is able to feel the earth's geomagnetic field. Among the many types of sensing that you should know took place then, and still do now, magnetoreception offers us a humbling experience as self-centred human beings.

Can you consciously feel the earth's magnetic field, let alone navigate using its energy; can you transduce it? At life's inception on earth (Act I scene 1), our ultimate universal common ancestor lived inside hydrothermal vents that "generously squirted" iron, being bombarded by the UV light that radiated intensely due to the lack of an ozone layer. Our ancestors survived by using powerful antioxidant iron-based protein system to scavenge intracellular free radicals generated by the synergetic combination of UV radiation and free form ferrous iron on early Earth. Their tremendous propensity left behind magnetic fossils that serve as priceless markers for reconstructing the earliest life conditions on earth.

This is all wonderful but what about magnetoreception, you might ask? Well, 2.7 billion years ago, as our common ancestors became motile, leaving the hydrothermal vents, they took advantage of their ability to metabolise iron as a mean to guide them; hence begun the evolution of magnetotaxis, based on a serendipitous opportunity. We can look at vision in a similar way: light transduction initially served the sole purposes of assisting in energy production and only later on as prokaryotes became mobile did phototransduction evolve into mediating phototaxis. One could even put it this way: transduction was co-opted into sensing as we define it, embracing the definition of "feeling". As they left hydrothermal vents and biofilm communities, bacteria started metabolizing iron as nano-sized, membrane-bounded magnetic crystals of magnetite (Fe_3O_4) and/or greigite (Fe_3S_4) called magnetosomes. Distributed as a straight single row inside the bacterium, magnetosomes establish a linear dipole that provides passive orientation as the bacterium swims actively by rotating its flagella. In conjunction with other responses (aerotaxis, phototaxis, chemotaxis, and voltage-sensing taxis), magnetotaxis guides the bacterium toward survival-promoting environments (Lin et al., 2019).

Act II, scene 6:

We are now at 1.85 billion years in the past. The first eukaryotes have evolved, finally. These larger single cells with nuclei, little organs (organelles) and a cytoskeleton, will take on the earth. They will share it with the prokaryotes and have fun. They will exchange DNA and even more with the prokaryotes. Much more. The little organs themselves likely originate from the engulfment of prokaryotes through phagocytosis (eating). At the party also are the newly evolved bacterial viruses (bacteriophages), assisting with DNA transfer. Oxygen levels continue to rise thanks to photosynthesis and the new eukaryotic contributors.

Transduction in eukaryotes favours multi-component systems. The ligand/receptor tends to be covalently bound to enzymes such as tyrosine kinase and phosphatases, which form second messengers including cyclic AMP (cAMP) and inositol triphosphate (IP3), which controls the release of intracellular calcium stores into the cytoplasm). New players, adaptor proteins, enable the interaction and the coordination of additional signaling proteins to specific stimuli.

Together, enzymes and adaptor proteins respond to multiple second messenger molecules. The level of complexity is significantly increased compared to prokaryotes. There are now cascades of phosphorylation, where individual kinases have the ability of phosphorylating numerous proteins until removal of their phosphoryl group. The larger volume of eukaryotes and their intricate intracellular organization (with specialized organelles, a cytoskeleton, and a nucleus) might have "*called for signal transduction systems with higher signal amplification and a greater number of components to bridge the spatial separation of the ligand-binding and DNA-binding domains*" (Cashin et al., 2006).

Accordingly, eukaryotes possess an even larger variety of signal transduction proteins than the already impressive array exemplified above by prokaryotes. In addition, signaling proteins are attached to a protein scaffold, ensuring their proper co-localization and isolation from other cascades within the large and complex intracellular space. Complexity brings more complexity. Sensory transduction tends to require longer phosphorylation cascades, adaptors, and scaffold proteins. There are also opportunities for 'cross talk' between different signalling pathways in eukaryotes. Beer is a great example; *Saccharomyces cerevisiae* (in Spanish, cerveza), a type of beer yeast, relies on crosstalk for mating. Yes, beer yeasts do have sex.

The first fossil record for sexual reproduction dates 1.2 billion years ago, about 100 million years after the first appearance of life on land, mushrooms (fungi). By 1 billion years ago, the first non-marine eukaryotes move onto land. They were photosynthetic and multicellular, indicating that plants evolved much earlier than originally thought. Underwater, eukaryotes remained unicellular; they experimented with new levels of structural complexity, such as for light perception.

A new organelle evolves: the ocelloid. First documented by Kofoid and Swezy (1921) in a few select planktons: at first glance, the ocelloid looks like a miniature eye self-contained within a single cell. Figure 22 depicts one such plankton, the *Nematodinium.*

Figure 22

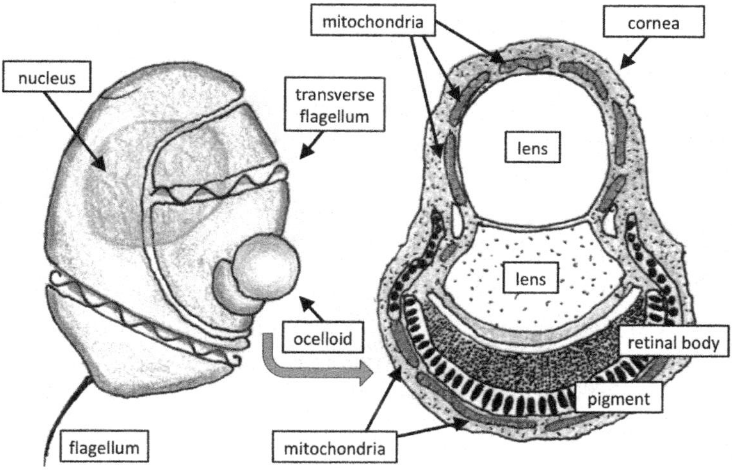

Distinctive landmarks of what we expect as an eye (based on ourselves) are obvious when observed at the microscopic level: a cornea, an iris, a lens, a retina and a pigmented layer (Francis, 1967). However, the ocelloid is not an eye, it is an organelle, each of its distinguishable feature is not made up of cells, but rather of mitochondria (for the cornea), polymers (for the lens), retinal-bound photopigment (for the retinal body) and brown pigmented bodies (thylakoids, for the pigmented epithelium and the iris), all of which originate from ancestral eukaryotes having engulfed prokaryotes (Gavelis et al., 2015).

Again, we intuitively grasp reality by looking at it. The analogies with a human eye are remarkable. It is much easier to be in awe with the feat of seeing what looks like an eye within a cell; much easier than the intricate intracellular pathways still under scrutiny in single cells, both for prokaryotes and eukaryotes. One can look at a schematic of molecules and links between them, but an image capturing structural complexity is in a way more "demagogic"; it easily appeals. My point is that complex diagrams are harder to grasp as one cohesive mental takeaway message. This being said, analogies with humans are palpable and in a way do deserve awe. Most of the multi-component pathways that have evolved, starting 1.85 billion year ago in eukaryotes – and if fact also even before 2 billion years for prokaryotes – are the foundation of our own physiology. Not much has changed and this fact in itself is nothing less than astounding.

Act II, scene 7

We are now at 1 billion years ago, unicellular algae thrive, forming complex structures, some of which will fossilize as multicellular and morphologically differentiated macrofossils. Findings by Tang et al. (2020) support that such fossils (from green algae, *Chlorophyta*) are in fact related to the ancestors of all the modern land plants we see today. These fossils could represent the first instance of multicellularity and cellular differentiation. The evolution of unicellular to multicellular species implies a paramount sacrifice from a cellular point of view. As part of an organism, single cells sacrifice their individual survival to the greater whole. As they specialize, cells need each other to survive as a whole. Under the pressure for multicellular organisms to survive and reproduce, nervous systems have evolved to sense the environment (internal and external) that entails threats and pro-survival sources.

At the same time, 1 billion years ago, some unicellular algae will acquire traits rendering compatibility for life on land (Strother et al 2011). These traits include antioxidant mechanisms to protect against damaging UV radiation while underwater, which became an asset for survival on land, where UV exposure was heightened.

From 0.85 to 0.63 billion years ago, plants consume CO_2 to the point of thwarting the atmospheric green house effect and thus cooling the earth. At the same time, plants also produce oxygen, leading to the formation of an ozone layer, eventually protecting them against the damaging UV radiation and further facilitating their land colonisation. By about 0.580 billion years ago, the Ediacara life forms (biota) represent large complex aquatic multicellular organisms. These elusive organisms left no descendants. All we have to study them are fossils, visually stunning fossils. Most macroscopic fossils are morphologically distinct from later life forms: they resemble discs, tubes, mud-filled bags or quilted mattresses. Palaeontologist Guy Narbonne suggested that these complex life forms might represent "failed experiments" in multicellular life, with later multicellular life evolving independently from unrelated single-celled organisms. Failed experiment or victim of drastic environmental changes? What if the mysterious Ediacara biota became extinct due to a fall in oxygen level? Had oxygen levels not lowered drastically to anoxic levels, as supported by increasing number of studies (Zhang et al 2018; Tostevin et al., 2019), animal life as we know it would have been altogether different, including us, human; a humbling reflection. What a fascinating period this would have been to witness; this real period and the hypothetic alternative of persistent Ediacara biota. In his book, "The Garden of Ediacara: Discovering the First Complex Life", palaeontologist Mark McMenamin articulates that Ediacara species Ediacarans “seem to have developed a central nervous system and a brain independent from animal evolution”. The possibility certainly exists, but we simply do not know.

It is time to leave speculations and embark on an epic journey in Act III, a feast for the eyes, where the next 0.5 billion years will see the evolution of all known multicellular species (animals). Buckle up here comes the Cambrian explosion.

Act III, scene 1

We are now at 0.542 billion years ago. Welcome to the Cambrian explosion. Multicellular evolution officially begins. The elusive multicellular Ediacaran organisms have been usurped; and so were the equally elusive first multicellular underwater Chlorophyta and the first land plants. Think about it, most (86%) of life's 3.8 billion years history is about single cells. The majority of evolution's trial and error involved single cells. Three billion, two hundred and fifty eight million (3,258,000,000) years during which single cells adapted to their ever-changing environments allowing them to pass on genetic information vertically and horizontally. But why did it take so long?

According to science writer Sam Kean (The Violinist's Thumb) bacteria evolved very slowly in part because only 2% of their energy was devoted to DNA copying, modification and maintenance; most of the energy went toward protein synthesis and locomotion. Everything changed when the engulfment of small bacteria by larger bacteria led to a unique symbiotic collaboration. These small bacteria, now living inside a larger cell, very slowly specialized through random mutations of their own DNA at a rhythm of one mutation every 3,500 years; no hurry. They eventually started making energy extremely efficiently using oxygen. Mitochondria had evolved.

We owe this knowledge to Lynn Margulis who had to face 15 rejections of her manuscript in the mid-sixties before... well, the rest as they say is history. Margulis was eventually vindicated; unfortunately her name did not quite make it into popular culture. Overwhelming proof comes from the fact that mitochondrial DNA is similar to that of bacteria in its sequence, storage form and ability to replicate in high number.

Host cells, now free to undergo more DNA replications and errors, acquired a rich repertoire of genes, many of which are required for cell adhesion, cell signalling and transcriptional regulation in modern animals. This is a non-visual awe moment for the reader: remember the mind-bugling diversity and intricacy of transduction pathways taking place in single cells. Host cells also had energy available to make novel organelles such as a nucleus. Eukaryotes had evolved. A big thank you for mitochondria and their very own bacterial-like DNA.

The eukaryote candidate precursors of multicellular organisms, are called holozoans, they include choanoflagellates, filastereans and ichthyosporeans (Sebé-Pedrós et al., 2017); however, environmental surveys support that countless potential additional candidates are out there, in marine and freshwater environments, having eluded scrutiny (del Campo and Ruiz-Trillo, 2013). Here is one of the many reasons to respect our planet: global warming threatens the very existence of many species, some of which we are still oblivious to and might hold pertinent keys to the conundrum of evolution to multicellularity.

The vast majority of all existing animal species (including human), their ancestors and extinct animal species, can be traced to fossils dating from the Cambrian Explosion. The oldest animal fossils belong to comb jellis (Ctenophora), sponges (Porifera), as well as corals and sea anemones (Anthozoa). How did these first multicellular species feel? We cannot realistically, within the context of this book, address all species having evolved during the Cambrian explosion. Let us use examples for the sake of comparing them with our own means of how we "feel'. Again here is a humble experience: looking at sensory pathways of species that roamed the earth over half a billion years ago, before all other animals had evolved.

Comb jellis come in diverse visually elegant shapes (figure 23). These mobile species rely on various sensory modalities, including vision, to guide them. They rely on an apical organ that harbours morphologically distinct ciliary photoreceptors (Horridge, 1964). Moreover, these photoreceptor cells (we are now multicellular) display the rich array of proteins of the phototransduction cascade. You will recognize the same players as described in the first chapter for human vision, such as: retinal-bound opsin, transducin, phosphodiesterase and cyclic GMP gated cationic channels (Schnitzler et al., 2012).

Figure 23

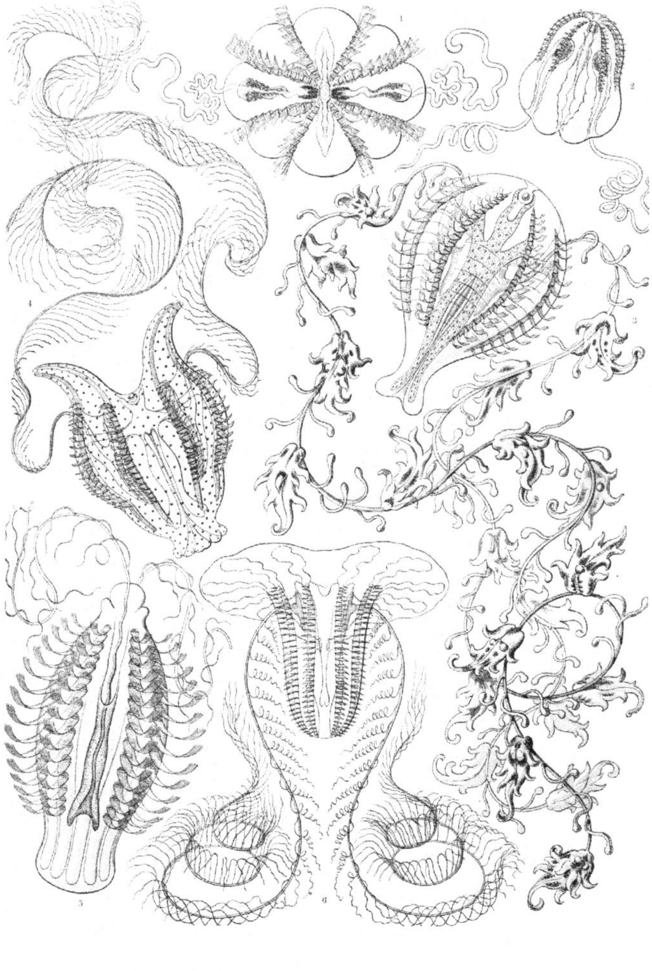

Comb jellis also have sensory organs for balance, similar to organs in our inner ears (otholith and semi-circular canals) that contain hair cells as mechanoreceptor mediating our sense of balance. The parallel with human is obvious. An organ situated at the opposite end from the mouth (aboral) comprises a balance sensor (statocyst) consisting of a statolith, a tiny grain of calcium carbonate, supported on four bundles of cilia, called "balancers", that sense its orientation. Instead of keeping the ctenophore automatically balanced (that is, with the statolith resting equally on all the balancers, following the gravitational force) its orientation is determined by specific behaviours such as capturing a prey where comb rows are put into reverse to spin the mouth towards the prey. This behaviour relies on concerted sensory inputs such as visual cues and gravity to dictate an output suited for survival: catching a prey.

While we know that sponges also perceive and respond to a wide variety of stimuli (light, chemicals, gravity, pressure, oxygen), the actual respective transduction pathways are still poorly understood; there is so much to be learned (Leys et al., 2019). For instance, sponges do not use opsins or any of the archetypal phototransduction players to respond to light. Furthermore, each sponge species (estimated between 5,000 to 10,000 types) responds using different transduction pathways.

As is the case with sponges, the sensory pathways of Anthozoa are also poorly understood. Their sensory system consists of simple nerve nets in the gastrodermis and epidermis, without any specialized sense organs other than the apical organ of the larval forms. Molecular characterization of this organ has revealed up to 78 candidate genes but the actual transduction pathways elude us at the moment (Sinigaglia et al., 2015).

Ultimately, the presumptive eukaryotic precursors (holozoans) or multicellular life and the first forms (comb jellis, sponges, corals and sea anemones) have set the foundation for the complex multi-component signalling pathways that are essential to multicellular life. The question "how do you feel", has further implication when considering multicellular species. Fertilized egg cells (asexual or sexual) must divide, leading to precursors of specialized cells in the right position and with precise timing. These cells must "feel" their environment in order to activate developmentally regulated genes.

All known animal species rely on seven fundamental pathways, entirely laid out over half a billion years ago (Babonis and Martindale, 2016; figure 24): 1) transforming growth factor b (TGF-b); 2) canonical WNT (cWNT) signalling; 3) nuclear receptors (NRs); 4) receptor tyrosine kinase (RTK) signalling; 5) Notch/Delta; 6) Hedgehog signalling; and 7) Janus kinase/signal transducer and activator of transcription (JAK/STAT). These vital pathways transduce extracellular signals into gene expression, regulating the target genes that determine cellular specialization. It is these pathways that established the division of labour, which resulted in the origin of novel cell types and the evolution of complex multicellular life.

The exact events underlying the evolution of the vast diversity of life forms represent a bustling area of investigation, with its associated debates and quest for funding. Basic research is not the flavour of the month, even though it might very well hold answers to potential therapies such as cancers, which involve the uncontrolled division of non-differentiated cells. Untangling the events that led to cellular specialization is bound to provide novel insights into preventing and treating cancers. In fact, justification of funding (with some success rates inferior to 10%) often involves invoking defined health-related benefits. Again, think of laser and GPS as examples of unpredictable basic research offshoots. Since we are mentioning cancer, one could view it as a vestigial natural tendency for cells to survive and divide, which when occurring out of context can kill the organism they are part of.

Figure 24

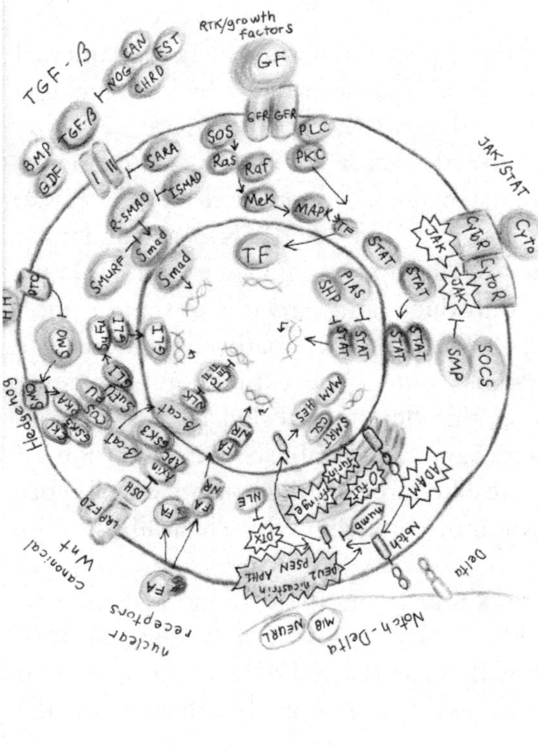

Act III, scene 2

The time is 0.535 billion years in the past. We are still early in the Cambrian explosion and numerous novel animal species have already started to evolve, mostly all in oceans: Chordates (underwater species), arthropods (trilobites, crustaceans), echinoderms, molluscs, brachiopods, foraminifers and radiolarians, etc.; each and every one displaying inventive sensory modes adapted to their lifestyles and environment and each and everyone warranting its own scrutiny but there is still so much ahead of us.

The chronological diary becomes increasingly detailed and relatively more precise, namely, at plus or minus one million years... The first footprints will appear on land. Many more new phyla will evolve: graptolites, crustaceans, cephalopods (nautiloids), chitons, species from the fossilization of the Burgess Shale, jellyfish, the first vertebrates with true bones (jawless fishes), conodonts, echinoids, and agnathan fishes.

At 0.444 billion years ago, the first of five major extinctions strikes. The Ordovician-silurian extinction eliminates 49–60% of marine genera and nearly 85% of marine species. However, marine life diversity rebounds with unrelenting re-flooding of continental shelves, which increases biodiversity within the surviving orders. New phyla evolve not only underwater but also on land: ray-finned fishes, trigonotarbid arachnids, land scorpions, fish with teeth, lichens, stoneworts, harvestmen, mites, hexapods, ammonids, tetrapods, and one of the earliest walking vertebrate: the long extinct Acanthostega (spiny roof).

After a relatively short respite of only 79 million years, a second major extinction strikes: the Devonian Extinction at 0.365 billion years ago. There is accumulating evidence that in fact there were as many as seven distinct events, spread over about 25 million years. Numerous tropical marine species became extinct. Overall, 19% of all families and 50% of all genera vanished. The reef-building organisms almost completely disappeared.

We made it to 0.363 billion years in the past, a time when the earth has a semblance of present day ecosystems. We are in the Carboniferous Period. The globe teems with insects, first on land and soon also air bound. Lush vegetation bares seeds that will birth majestic forests and dense jungles. The oceans welcome the fiercest of all predators, sharks, placed the highest on the food chain's podium. Four legged animals (tetrapod) walk on land. By 0.320 billion years, precursors to mammals (synapsids) and reptiles (sauropsids) branch out.

Skipping details on new phyla, we arrive exactly at 0.251 billion years ago, a landmark moment. The majority of marine species are annihilated, an event known as the Permian–Triassic extinction (third of the five major extinctions), which paradoxically, is followed by the remarkable flourishing and diversification of marine life known as the Mesozoic Marine Revolution.

Having more predators in the oceans implies that there are more opportunities for adaptation with regard to sensing and reacting to the environment: distinguishing friends from foes. Meanwhile on land, the first dinosaurs evolve. We are at 0.225 billion years; the majestic herbivorous prosauropods thrive in the lush forests. Their impressive dimensions house the largest guts to have existed. Gargantuan guts are required to digest the luxurious albeit nutrient-poor plants they graze.

At the same time in what is now western Texas, the first mammals (genus Adelobasileus) evolve in a precarious environment dominated by the predatory dinosaurs *Coelophysis*, but also at a precarious time because yet again another considerable extinction looms ahead. Two hundred and ten million years ago, the Triassic/Jurassic extinction (fourth of the five major extinctions) wipes out numerous terrestrial vertebrates, allowing countless new dinosaurs to evolve; these will epitomize the "Jurassic" period. Yes, Tyrannosaurus rex is one of them.

Act III, scene 3

Dinosaurs reign the earth from 225 to 66 million years in the past, the time of yet another major extinction, the Cretaceous–Palaeogene extinction (last of the five major extinctions), which eradicates all of them. Actually, over half of all animal species on earth vanish. Birds are spared. Mammals are now the dominant species, counting among them predators such as the carnivorous creodonts and miacids as well as primates.

We are now 60 million years in the past. The very first primates (plesiadapiformes) have just evolved from the small terrestrial mammals (Primatomorpha) and now live on trees in tropical forests. Adaptations to these prehistoric environments have determined the unique features of primates, which as members of this club, we share. We owe our large brains, visual acuity, colour vision, altered shoulder girdle, and dextrous hands, to the requirements of having to survive and reproduce on the trees in 60 million year old tropical forests.

As chronocentric animals, it is counter-intuitive that the features typifying primates would have evolved prior to that of many other species such as tapir, rhinoceroses, camels, modern-type butterflies, moths, bears, pigs, and cats whose oldest fossils are at least 10 million years younger than those of the first primates. One could rationalize that it took many more years for primates to evolve as the first humans, the hominids. True, we must wait until 6.5 million years ago for evidence of the first hominine. *Sahelanthropus tchadensis* lived in a region that is now known as the Djurab desert, in Northern Chad. However, many other species had yet to evolve, still.

Five million years ago welcomes tree sloths, hippopotamuses, zebras, elephants, lions, ancestral dogs, kangaroos, leopard and seals. An obstinate chronocentric *Homo sapiens* could say, yes but how about real humans, surely they evolved much later than these "inferior species", human must be at the top of the top of evolution's master plan, right? Well, there is no "master plan" anyhow. Dear obstinate chronocentric human, do you mean Stone Age human who used tools, the genus *Homo habilis*? The first fossil records date 2 million years ago, leaving ample time for many more so-called "inferior species" to evolve. We mentioned above that mitochondrial DNA mutates at a rate of one event per every 3,500 years. What we failed to cover is that mitochondrial DNA is only passed from mothers to their progeny.

Knowing how many mutations occurred serves as a clock. In this manner, scientists estimate that all humans living on earth today descend from a single female *Homo sapiens* who lived in Africa 170,000 years ago; you can call her "Mitochondrial Eve". However, there were many other females, she just happens to be the oldest matrilineal ancestors of all humans living today (Kean, 2012).

I think we can stop our history of the senses here. Of course the concepts of superior or inferior are dangerously flawed. Eugenics is a loaded gun pointed not only at humanity but also at the whole world we share with all life forms and the reality we can and cannot grasp. Just to put a nail in the coffin of anthropocentric views, let us remember that at this very moment, bacteria and viruses are still evolving and rather swiftly so. I am writing this book during the full-blown COVID-19 viral pandemics, in social isolation.

The year is 2021. We are forcefully reminded that science grasps a reality we cannot escape as a society. Bacteria and viruses are evolving at a much faster pace than we are, genetically speaking. They are everywhere on earth:

1) On its surface (a tea spoon of soil can contain up to a billion bacteria with more than 10,000 species, mostly uncharacterized, and up to 1.2 trillion viruses; the estimated number of bacteria covering the earth is equivalent to the number 2.55 followed by 29 zeros).

2) In the ocean (a single teaspoonful of seawater contains over 100 million viruses, and the cumulative microbial mass is estimated to reach up to 23 billion tons of carbon).

3) Up to 10,000 meters high in the atmosphere (over 2,100 different microbial species have been estimated to populate clouds).

4) Let us not forget the ones living inside and outside animals (the human microbiome represents 3.8 trillion bacteria with over 10 thousand different species and even more viruses: 380 trillion).

Forming such ubiquitous populations and being able to exchange genetic material, no wonder why viruses and bacteria can evolve extremely rapidly. Humans, in comparison, undergo a much slower and less remarkable genetic evolution. If we look very hard, we can find two genes, ASPM (abnormal spindle-like microcephaly associated) and Microcephalin (MCPH1), that have continued to evolve adaptively since the emergence of anatomically modern humans, with the adaptive alleles rising from one copy, as recently as ~6,000 and ~40,000 years ago, to a worldwide frequency of ~30% and ~70%, respectively (Mekel-Bobrov et al., 2005; Evans et al., 2005). However, it is unclear as to what advantage these genes might provide; both are implicated in a deleterious condition, microcephaly (smaller brain size). On the positive side, perhaps, a role for ASPM has been established with higher-level brain functions in the temporal cortex related to human communication (Wong et al., 2012).

Whether certain genetically determined traits are adaptative depends on social acceptation, and integration. In his book "Neurotribes", Steve Silberman eloquently reminds us that mutations associated with autism are not an error of nature but the result of natural variations in the human genome. Silberman highlights the unusual skills and aptitude offered by these mutations, as a blatant example that evolution is still taking its course. Our responsibility as a society is to embrace differences and provide the learning resources that respect and foster such "neurodiversity".

Here are a few well-known people, some of whom you might not know, who display traits that fall within the autism spectrum: composer-singer David Byrne, actor Dan Aykroyd, actress Darryl Hannah, singer Courtney Love, actor Sir Anthony Hopkins, astronomer Carl Sagan, animal science professor Temple Grandin, autism researcher Michelle Dawson, footballer Lionel Messi, artist Vincent Van Gogh, scientist Albert Einstein, scientist Isaac Newton, inventor Nicolas Tesla, scientist Charles Darwin, and as a final example: software revolution leader and one of the richest man in the world, Bill Gates.

Unfortunately, still today, autism spectrum is clinically considered as a developmental "mental disorder" instead of a different form of neural processing, or using Steve Silberman's terminology: "neurodiversity". Perhaps it is the environment that has to adapt. Society has to be educated to the beauty of "neurodiversity". There is so much to learn on how sensory information is processed and what outcome is produced. Autism spectrum and other forms of diversities, such as dyslexia and ADHD, provide an infinite array of potentials, which we just have begun to recognize. We still have a long way to go: ignorance is a major hurdle. The answer to "neurodiversity", which should be praised and cultivated, lies between sensors and effectors. The category of autism spectrum is associated with diversity in sensing and reacting. It must be considered not as a defect but rather as a difference, a fascinating difference.

Chapter III

Between Sensors and Effectors

Between sensors and effectors, the in-between becomes sense, trying to make sense, compelled to make sense. Let go, breathe, and "come back to your senses".

The point of recounting the timeline of how we evolved, was to provide a broader context of how we "sense", of what it means to "sense", to "feel", to "think", to "have a soul". However, trying to pinpoint an exact moment when consciousness evolved or to distinguish which species are conscious and which are not is a trap set by rigid definitions (Feinberg and Mallatt, 2017). Cusp to compliance is a moving nebulous target. In this chapter, we will focus on the linkage between sensors and effectors, in the context of evolution and development.

Let us appreciate the continuous process of evolution, a continuity that is also characteristic of embryonic development. We mentioned the 7 basic transduction pathways underlying animal development from a fertilized egg (single cell, ovum). One cannot pinpoint an exact moment when consciousness arose in evolution and when it arises during embryonic development. In humans, synchronized cellular activity (oscillations) between the reticular formation, the thalamus and the prefrontal cortex is certainly involved in consciousness (Crick, 1984), but nothing is all or none. Think about it consciously: while fully awake, you are not always-uninterruptedly conscious or aware or responsive to sensory stimuli. Think about ocular saccades; try again to catch your eyes moving when looking at yourself in a mirror. Think about all that you do automatically beyond what your autonomous nervous system is already accomplishing.

An ethical question arises. When does an embryo become conscious? Which species are conscious? Other species than human feel to. Undeniably, we have moved on from philosopher René Descartes's statement that only human truly feel (are sentient) whereas other life forms are no different than any objects. But we still have a long way to go as a society; the resurgence of aboriginals' relationship with nature is welcomed; we are one with the trees (remember we share the same last ultimate common ancestor, LUCA). Yes I am digressing; you get the point. Let us embark on yet another story: how sensors and effectors are linked in different levels.

Level 1: In unicellular species, sensory receptors are linked with effectors via transduction pathways, which can be regrouped as one-, two- or multi-components, or extra-cytoplasmic factors, as reviewed in the second chapter. Cells that constitute multicellular species also can display sensing and effector standalone properties, across evolution, from jellyfish to human. During the larval stage of box jellyfish (*Cnidarian cubozoan*), phototaxis is mediated by multi-functional single cells that comprise photoreceptor, shading pigment and a locomotor cilium (Nordström et al., 2003). In humans, multi-functional single cells such as in guts and lungs can "feel" and act by beating their cilia, the outcome is the movement of food and the clearance of mucus, respectively.

In addition, there are instances where unicellular species establish junctions between themselves, producing a coordinated output in response to specific stimuli; we can call this "level 1 plus". Such occurrences are characteristic of a colony formed by unicellular species. Remember our eukaryote candidate as the potential precursor of multicellular species: the choanoflagellates? A novel species identified in 2019, *Choanoeca flexa*, forms cup-shaped colonies that can rapidly invert their curvature in response to changing light levels, morphing the unified mass of unicellular species from slow (flagella-in) to fast (flagella-out) swimming (Brunet et al 2019). In fact, the two other closest unicellular relatives of metazoans, ichthyosporeans and filastereans, also can establish colonies. Gene expression studies in all three groups revealed that cells synthesized synaptic proteins when regrouped as colonies. The presence of synaptic proteins (such as Dlg/PSD-95, Homer, Shank, Synaptophysin, Synaptogyrin, and Complexin), in unicellular species surviving as a colony, preceded the evolution of multicellular species and nerve cells, let alone the synapses that connect them (Burkhardt et al., 2017).

What we can take home here is the very foundation of nervous system evolution: the coordination of cells that allows them to move as a whole. Hold this thought for the fourth level, when we first introduce interneurons. Spoiler alert: in their simplest form, interneurons are coordinators of muscle-driven movement.

Level 2: In multicellular species, in their simplest form, sensory cells can also act as motoneurons by directly contacting effector cells. Dual roles have been documented in nematodes for two neurons, IL1 and IL2, known to act as both sensors (mechanoreceptors and chemoreceptors, respectively) as well as motoneurons (White et al., 1986). IL1 mediates nose-touch responses and foraging (Hart et al., 1995) while IL2 has been associated with nictation, a fancy name for when a worm stands on its tail and waves its head in three dimensions (Lee et al., 2011).

In the plankton *Platynereis dumerilii* (Jékely et al., 2008), rhabdomeric photoreceptors (as opposed to ciliated as in human) directly synapse onto locomotor ciliary cells. The outcome is nothing less than the largest biomass transport on earth: a light-driven movement of plankton.

Level 3. At this next level of complexity, a sensory cell is linked to an effector cell by a third cell. This linking cell is not an interneuron. Here is the reason: sensors are neurons, but effectors are not. Typically, effectors are muscle cells. Therefore, these third cells inherited the name "motoneuron". The contact between motoneurons and effector cells bares the more generic and widely encompassing name "neuroeffector junction". The junction that gave its name to "motoneurons" is the neuromuscular junction; it is always excitatory.

Other types of junctions exist, close to home: they are part of our autonomic nervous system (unconscious actions). These can be either excitatory or inhibitory. Examples include gut motoneurons innervating smooth muscle, cardiac muscle, and glands. Unlike synapses, neuroeffector junctions are juxtapositions of motoneuron nerve terminals with effector cells that are open to the extracellular space, allowing for various levels of separation and area coverage between the release sites and the effector. They are the sensors eliciting outputs that maintain your body's homeostasis, without any conscious knowledge, awareness. The sensors detect temperature, oxygen, carbon dioxide, acidity, osmolarity, glucose levels, hormone levels and any other biomarkers that are indicators of your general body status. Effectors provide the necessary physiological actions to maintain your general body health. These complex tasks are very much the responsibility of the autonomic nervous system.

Another way to think about this third level is to consider all the cells in multicellular species, including human, that have sensory receptors and release diffusible factors that bind to remotely located neuroendocrine cells, which release hormones in the blood stream that will act on effectors cells throughout the body.

In the brain, neuroendocrine cells are regrouped in the hypothalamus, where they release factors in the bloodstream (such as oxytocin and vasopressin) and in the anterior pituitary (releasing hormones). These factors regulate functions in the entire body, encompassing stress responses, thermoregulation, sleep, and reproduction.

Other neuroendocrine cells known as chromaffin cells are found on top of the kidneys in the adrenal medulla, where they release adrenaline into the bloodstream under the control of the sympathetic division of the autonomic nervous system.

In this very context, we can include any nervous system cells, even interneurons themselves. But at this level, these cells act as sensors. At the risk of repeating myself again and again in this book, interneurons themselves do not contain sensory receptors eliciting conscious precepts, or awareness. Yes, interneurons do contain receptors that sense and respond to their environment, but their response is to release molecules, neuromodulators, which can act at a distance on neuroendocrine cells.

Examples of neuromodulators secreted by interneurons include: cholecystokinin, somatostatin, neuropeptide Y, galanin, and vasoactive intestinal peptide. Of interest, imbalances in any of these neuromodulators have been linked with various psychiatric disorders (Marín, 2012). Most of the sensing activity in interneurons occurs during development, where they play elemental roles in neural circuitry differentiation and functional refinement (Fishell and Kepecs, 2019).

Upon CNS maturation, glial cells – the main culprits in maintaining homeostatic balance (nicknamed "housekeepers") – achieve most of the sensing inside the brain itself. They also act as effectors themselves; therefore they belong to level 1: the same glial cell senses and acts. Glial cells sense a wide range of biomarkers such as osmolarity, acidity, volume, oxygen, carbon dioxide, and glucose levels. Their outputs include pumps, transporters, channels, and neurotransmitter uptake receptors, all of which correct homeostatic deviations detected from the sensors.

Finally, here is one last example of level 3, close to home, this time for your body's balance: the knee-jerk reflex, which contributes to your ability to stand up. A precise hit (with a pointy rubber hammer) just below the kneecap, targeting the patellar tendon, activates muscle spindle mechanoreceptors of the quadriceps, which directly synapse with alpha motoneurons in the ventral spinal cord, in turn activating the quadriceps femoris muscle and resulting in an upward kick of the leg. To actually work, the reflex requires coordinating the simultaneous relaxation of an antagonistic muscle, the flexor hamstring muscle. This is where interneurons are required. At their very foundation, interneurons are inhibitory and play a role in coordinating outputs (effectors); in the case here: muscles. We are ready to embark on the next level: interneurons as linkers.

Level 4. Finally, interneurons come into play as linkers between sensors and effectors. At this level, they mediate the coordination of muscles. The knee-jerk reflex's upward kick requires mechanoreceptors to simultaneously synapse with: 1) an alpha-motoneuron that activates an extensor muscle (quadriceps femoris); and 2) an interneuron that inhibits the alpha-motoneuron, which contacts the antagonistic flexor hamstring muscle, resulting in its relaxation. At this level, sensors (input) and effectors (output) are separated by at least one interneuron and a motoneuron.

One can imagine almost limitless combinations of sensors, interneurons, motoneurons and effectors (muscle cells), leaning themselves to sophisticated patterns of muscle activation and inactivation. Ultimately, the evolution of interneurons made muscle coordination possible, allowing organisms to navigate their environment (Keijzer et al. 2013). Only later on, did nervous systems evolve to act as emulators and predictors of the external world, which deserves a chapter of its own, the fourth chapter, devoted to the next and last level.

Oldest examples of this fourth level are found in the Platynereis group of planktons with the startle response. Mechanoreceptor cells activated by water-borne vibrations, contact interneurons which then contact motoneurons that connect with ciliary cells. The outcome is a closure of the cilia and an activation of muscle cells. These effector cells are of an organ called parapodia, which protects the species from predators (Bezares-Calderon et al 2018). Here is another instance where evolution does not only lead to higher complexity. The knee-jerk reflex pathway is the essence of simplicity when compared to the already evolved complex neural circuit mediating startled responses in planktons (for a schematic of the circuit, see figure 6b from Bezares-Calderon et al., 2018). Simple reflexes are evolutionarily preserved when they contribute to the survival and reproduction of an organism. In parallel, more complex circuits evolve. All levels of complexity coexist in one organism such as in us humans.

Here is another example of a complex reflex: the rapid tail flipping that produces powerful swimming strokes, thrusting crustaceans backwards away from danger. Also named "lobstering" or technically the "caridoid escape reaction", this innate escape mechanism is a common lifesaver to marine and freshwater crustaceans such as lobsters, krill, shrimps and crayfishes. Numerous interneurons are involved in coordinating the muscles producing the brisk abdominal flexions. Therefore, interneurons do not only link sensors with effectors but can orchestrate complex behaviours, in some instances, without sensory inputs First described in 1946 by C. A. G. Wiersma who studied crayfishes" giant interneurons of the tail. He showed that these interneurons are central players. They "command" the escape reaction. Henceforth, Wiersma referred to them as "command neurons" in a 1964 publication (Wiersma and Ikeda, 1964). His landmark findings established that the activation (depolarization) of specific neurons could precipitate complex innate behaviours not only in crustaceans but also in several other species. Of note, command decisions are increasingly seen as being governed by networks of interacting and redundant cells, not simply by single neurons.

Elaborating further on the role of interneurons in coordinating muscles, we must talk about central pattern generators (CPGs): the neural circuits producing the patterns of neural activity that generate rhythmic motor behaviours such as swimming, crawling, walking, feeding (chewing, swallowing) and breathing. These groups of interneurons can function independently of sensory feedbacks or of any other patterned inputs. Their basic organization is similar between a vast range of species including nematodes, slugs, snails, zebrafish, lampreys, tadpoles, reptiles, birds or any mammals such as humans.

Already proposed by Thomas Graham Brown in 1911, CPGs that allow the whole body to "wave left and right" (as in swimming and crawling) consist of a half-centre oscillator, in which two mutually inhibitory 'halves' alternate in activity. We owe a great deal of our knowledge on CPG networks to Sten Grillner's studies of axial muscle-driven motor circuits in the lamprey (Grillner and Manira, 2020). Grillner and his colleagues defined the core neuronal constituents of CPGs. Adding limbs to body movement, the CPGs include oscillators for flexors and extensors. Furthermore, adding fins in fishes and wings in birds requires even more complex patterning than simple left–right alternation (reviewed in Katz, 2016).

Before we move to the final level where "the soul" lies – can you wait until then? – I would like to share observations about CPGs for locomotion. Studies in laboratory rats showed that complete mid-thoracic spinal cord lesions at day 7 postnatal are followed by the ability of the animals to walk as adult (Commissiong and Sauvé, 1993). However, similar interventions one week later in development lead to complete paralysis. Comparison between both developmental levels reveals that corticospinal axons, which mediate conscious motor actions, have not yet reached the mid-thoracic level at one week of age. Of note, 7 days cordotomized rats can only walk providing that hindlimbs are stimulated by a pinch or a direct contact with the ground. I should also add that the lesions, independently of the intervention age, cause the loss of bladder reflexes. To ensure the survival of the experimental animals, the researcher must manually massage the bladder to express urine; otherwise the bladder would fatally burst.

Talking about repelling views, picture the decapitation of a chicken; it starts running without its head. I admit the picture is sickening. For whoever ate chicken, I am sorry to say this event must have happened so you could enjoy your meal. A digression for you: if you have not read it yet, consider Jonathan Safran Foer's book "Eating Animals" for a more encompassing (compassionate) comprehension of what eating meat implicates. No moral judgement here, just a reflexion. Back to the headless chicken; it runs and flaps its wings precisely because CPGs are freed from their corticospinal control. As long as there is oxygen and energy available the chicken will generate rhythmic limb movements.

In healthy rodents, and also in humans, CPGs for locomotion are activated by will, i.e. through corticospinal neurons. When struggling to develop treatments for paraplegia and quadriplegia caused by spinal cord trauma, one is faced with the fact that CPGs might be intact but not amenable to activation. A promising avenue is the use of functional electrical stimulations below the spinal cord lesion sites (Luo et al., 2020).

One very last thing about CPGs in birds, they are also involved in generating birdsong. It is estimated that out of 10,000 known bird species, over 4,000 of them sing. They do so for multiple purposes such as courtship, rivalry, nest building and prey capture. Young birds learn their songs by listening to a tutor, in a manner similar to how humans learn to speak. We can study how we speak, in term of the neural basis, by using birds as models, and comparing them with each other. Two forebrain regions are involved in producing birdsongs: the high vocal centre projects to the region robust nucleus of the archistriatum which in turn projects to the brainstem regions that control the syrinx and the respiratory systems using CPGs. Complex motor outputs for the generation of unique birdsongs (including stereotypical, non-motor, non-rhythmic neuronal activity sequences) can be generated by CPGs that are themselves contained inside larger CPGs: 'nested' CPGs (Berkowitz, 2019).

In humans, strictly speaking, speaking is not generated by CPGs, at least not in the spinal cord and brainstem where they lie. Rather, there might exist analogous assemblies of neurons in our cortex that do generate patterns related to vocalization; these are referred to as cortical CPGs. I just wanted to say that, loudly, so you could hear this. Recent evidence supports the existence of cortical CPGs, not for vocalization (not yet) but for finely tuned arm movements, which rely on sensory input via the thalamus (Sauerbrei et al., 2020).

You might not be so inclined to admire the musical quality and complexity of sounds emerging from other species such as frogs, but here is another promising approach to understanding vocalization at many levels, from physiology, to behaviour, and from development to evolution (Kelley et al., 2020). Tadpoles (*Xenopus laevis*) are amenable to CPG isolation and manipulation as a viable whole brain preparation in a dish (*in vitro*). For instance, *Xenopus* frogs have six types of calls, one of them, the advertisement call, is generated by two CPGs distributed between the anterior and posterior hindbrain (parabrachial nucleus and nucleus ambiguus) on each side. The left and the right CPGs are coordinated by intricate interneuron projections between the anterior and the posterior nuclei. A descending input from the amygdala coordinates the simultaneous initiation of advertisement calling. Without this input, vocal responses of male to female calls become socially inappropriate. Here we have a reductionist gem, a link between precise neural activity and social behaviour.

Findings linking individual interneurons and distinct behaviours increasingly populate scientific literature. Ancient species, such as gastropods (snail and slugs), allow tracking the evolution of individual interneurons in a neural circuit alongside behavioural adaptation (including moving and feeding) to known shifts in environmental factors. Tracking is achieved by establishing their total gene expression patterns combined with anatomical, pharmacological, and electrophysiological methods, all that for a single interneuron that has its own ID number (Benjamin and Crossley, 2020). With their distinct identities, we can now follow single interneurons and their related behaviours throughout evolution, at least in so-called simple and ancient species such as gastropods. We owe much of the knowledge and tools to South African developmental biologist Sydney Brenner, who introduced the archetypical model *Caenorhabditis elegans* (nematode) in 1963. Brenner published a seminal paper in 1986 on the exhaustive characterization of *C elegans*'s 302 neurons, falling into 118 classes (White et al., 1986). Not only does each cell have their own ID number (https://www.wormatlas.org/index.html), but all circuits and connections (chemical and electrical synapses) have been defined (https://www.wormwiring.org). There are over 5000 chemical synapses, 2000 neuromuscular junctions and 600 gap junctions.

The gigantesque monastic feat of mapping an entire nervous system has only been repeated once, in the tadpole larva of *Ciona intestinalis* (Ryan et al., 2016). Using 7000 consecutive electron microscopic sections (each one being 10 nanometres thick), the lab of Ian A Meinertzhagen at Dalhousie University identified 177 neurons forming 6618 synapses including 1772 neuromuscular junctions and 1206 gap junctions. This is fundamental research, it is not glamorous, but the implications are fundamental (no pun intended). Brenner would have been proud; unfortunately he did not live to fully realize the ramifications and extent of his legacy.

Why not establishing a wiring diagram (connectome) for the human brain? Size is the foremost hurdle. Even a mouse brain is about one hundred thousand times larger in volume than that of the nematode. The limiting factor is the absolute requirement to rely on 10 nanometres thick section for electron microscopy imaging. Nevertheless, we can extrapolate from these two invaluable connectomes. In terms of how the nervous system develops from a single fertilized cell, Brenner put forward two plans. In the first, the function of cells is determined by its genetic lineage. In the other, its function is determined by the function of its neighbours after cell migration. Current findings show that most species (not only nematodes) follow various levels of combination of these two plans to transfer information to new cells during development. Early in development, interneurons act as sensors, guides, support cells, and activity coordinators. Later on, as the cortex achieves developmental maturation, interneurons become active contributors of reality production.

To the infinite and beyond (a teaser to level 5). While interneurons evolved as coordinators of muscle activation, as an offshoot (also called exaptation) they formed intrinsically active networks that generate a multisensory predictive interpretation of the outside world, which led to a further offshoot: a prediction independent of any sensory inputs, that we know as "feelings", "thoughts", even "the soul". You can experience sensing, you can sense, you can feel, you can do all of that without input from sensory organs. When doing so, you are no longer grounded in the reality (space and time) that you have evolved to produce. You can ruminate without limits. This is perhaps why paying attention to your breath can ground you, calm you; doing so brings back the evolutionary roots of linking sensory inputs with outputs. However, when trying to do so, naturally your thoughts will drift away as it is their very nature to be active independently of sensory inputs. Your thoughts will drift away a thousand times while you aim to solely focus on your breath. It takes a great deal of patience and training to continue this exercise.

At the risk of repeating myself, interneurons do not themselves possess any sensory receptors that elicit conscious perceptions; we are oblivious to their activity. We do not "feel" interneurons. The generation of conscious perceptions does not entail generating a perception of the interneurons that generate this percept; otherwise, it would create a vortex. An infinite loop ensues when attempting to use thoughts to catch thoughts, a situation described in the book "I of the Vortex" by neurophysiologist Rodolfo Llinás. The thought about thoughts catching thoughts has been around for some time. As if we could ever succeed to catch thoughts with thoughts...

Philosophers are experts at this approach; they come up with numerous schools of thought.

"The knower and that which he knows are both one, and he who unites and that with which he unites are one, and seer and seen are one." Abu Bakr Ibn al-'Arabi (1076 - 1148)

Breathe...

The last level, level 5, is about the generation of the reality we experience, we are ready for the next chapter.

Chapter IV

Our Brain Generates the Reality we Experience

"*When we look at a rock what we are seeing is not the rock, but the effect of the rock upon us.*" Bertrand Russell

If you look at a rock and see it, it is because your brain generates its image, not like a camera but rather by building it from scratch. In the first chapter, we used vision as an example of how the brain constructs the reality we experience based on sensory inputs. We went over the multiple changes involved in ultimately seeing. Shall I reiterate them here, briefly while attempting to avoid scientific jargons? Before you can even look at the rock, your eyes move, and perhaps your head as well, so that light bouncing off the rock can enter inside your eyes. These movements are part of the effect the rock has on you. Once you are looking directly at the rock, your eyes make minuscule movements so that light emerging from the contours and salient features are focused on a tiny area at the back of your eye, the fovea. It is the act of looking at the rock that causes these miniature eye movements. Light originating from the rock will change the shape of small molecules, which themselves will change the shape of larger molecules, which will in turn kick start multiple chemical and electrical changes inside your eyes. A series of electrical impulses will then be sent to your brain along your optic nerve, modulating its activity. Information about specific aspects of the rock's feature, its location, its fine details and its various shades of colour, will be sent separately to different areas of the brain to be merged only later into a coherent image. Up to 50 different brain areas can be involved in the process of looking at the rock. Avoiding jargon beyond this point is a task I am not up to. Feel free to go back to the first chapter. The take-home message is that looking at the rock elicits complex events in your brain. The operating word is "elicit"; therefore, Bertrand Russell had the insight to understand this fundamental concept.

Do allow me one science word though: interneurons. Ultimately, interneurons – the neurons between sensors and effectors – are responsible for generating the reality you perceive. You project the experience of a rock as occurring outside your brain. You look at the rock and take it for granted that it is there, immutable and unalterable, you do not need to deliberately activate each of the many processes that take place in generating its image. What a relief, otherwise I predict severe headaches.

If you do experience severe headaches from the hangover after having enjoyed fermented libations with your new cocktail party friend – whom you shared that we are blind on average 4.5 hours of our waking days – you might see two rocks, or just a single blurry one. Or if you took hallucinogenic drugs, the rock might change colour or transform into a monster or even a cathedral.

Your brain determines what you see. You can reinforce the illusion by touching the rock. Again, touching will induce changes in brain activity that will be synchronized with the creation of tis image using multisensory integration. Bertrand Russell could have said: When we touch a rock what we are feeling is not the rock, but the effect of the rock upon our brain. I liberally switched to "brain" instead of "us", to avoid the confusion that the rock touches our body directly, but not our brain. We could devote an entire chapter on all the unconscious events that influence how we define the rock in its context. Some players to consider: 1) its distance evaluated by stereoscopic vision; 2) by moving our body to induce what is called a parallax; 3) by counting how many steps are required to move towards it; 4) by noting that its perceived size changes as you get closer; 5) by the presence of glare or not; 6) by its colour motifs; 7) by its shadow and the shadows casted by nearby trees on the rock; 8) by noting the location of the sun; 9) by experiencing the memory of having seen this rock before and how you might have felt then, etc. When you think of all the cerebral activity generated by the rock, you realize (here is a cheap shot) that the rock truly rocks.

> While interneurons evolved as coordinators of muscle activation, as an offshoot (also called exaptation) they formed intrinsically active networks that generate a multisensory predictive interpretation of the outside world, which led to a further offshoot: a prediction that is independent of any sensory inputs, which we know as "feelings", "thoughts", "self", even "the soul".

Interneurons have no sensory receptors that lead to conscious experience; the self is the same as a predictive mental model, it is part of it, it is it. The use of words and terms are a matter of intense debate and contention (Summers and Anson, 2009). You might say that the brain has anatomically distinct areas devoted to respective tasks such as motor (action) and premotor function (planning the action). If we consider interneurons in the motor cortex alone, studies in rodents have shown roles in sensory integration, behavioural strategizing, working memory, and decision-making (Ebbesen et al., 2018). Therefore, not only definitions can vary between terms but also areas baring specific names. The use of neutral terms such as frontal cortex, instead of motor, might be more encompassing when it comes to defining functions. As Anil Seth stated in his succinct and to the point book "30 Second Brain": "*Cognitive functions are neither fully localized nor wholly distributed in the brain - each function depends on a complex but specific network of interacting brain regions*" (Seth, 2017).

The concept here is that interneurons interpose themselves between sensors and effectors. Thus, we are oblivious to the activity of interneurons. However, we have developed tools to study them, to "sense" them. As we discussed before, vision is the dominating sense when it comes to discovering our physical world. No exception here, we "sense" these senseless interneurons by depicting them visually. Some interneurons are particularly, in fact stunningly, photogenic.

One of the first to reveal interneurons visually was Camillo Golgi, in 1873, who relied on his silver impregnation method. Golgi hardened thin slices of tissue in potassium bichromate and ammonia, to then immerse them in a silver nitrate solution. At that time, Golgi considered the stunning images of his "reazione nera" (black reaction) as representing "diffuse nerve networks" (Golgi, 1873), as opposed to individual cells. As hard as it might seem to believe today, nobody had realized the implications of his discovery.

By chance, after 14 years had elapsed, neuroanatomist Ramon y Cajal (May 1st 1852 - Oct 17th 1934) visited psychiatrist Luis Simarro's house in Madrid, where he first saw a Golgi impregnated preparation (tweaked by Simarro) with his very own eyes and the assistance of a microscope (Cajal, 1917). It was love at first sight. In a makeshift laboratory set up in his kitchen, Cajal hunted for cells in "the garden of the grey matter" (Cajal, 1917). The operating word is "cells" as opposed to network or reticulum. It was Cajal who confirmed the "Neuron Doctrine" by establishing that nervous tissues are composed of individual neurons, each having distinct anatomical features. He characterized thousands of different neurons and their networks across a wide range of species, developmental stages, and diseases (Cajal, 1909-1911). Cajal was on fire.

Contrasting jellyfish and comb jellis (where he described two types of neurons, sensory neurons and motoneurons) with worms, Cajal noted a third class of neurons. He wrote: "*But another element complicates it, and its mere appearance marks all the progress, enormous, all the superiority of the worms over the coelenterates. This new element is the intermediate or association neuron. With the association neuron, the multicellular compound truly becomes an animal. Sensitive arousal, however localized it is in a point of the skin, is no longer confined; it does not only bring about the reaction of the muscle fibres located in its zone of influence; all or part of the other fibres are also set in motion, depending on the intensity of the excitation, and the whole animal can, if necessary, vibrate and revolt at the slightest shock*". (English translation by the author, from French in Cajal, 1909-1911).

With his astute description of "intermediate or association neuron", Cajal had anticipated the relevance of interneurons. He poetically suggested that interneurons were "... *the mysterious butterflies of the soul… whose beating of wings may one day reveal to us the secrets of the mind*" (English translation by the author, from French in Cajal, 1917). Referring to the work of an equally insightful Italian psychiatrist, Ernesto Lugaro (1891), Cajal wrote: "*Thus, according to Lugaro, the intellectual process is linked to the existence of interneuron connections.*" Cajal added more visionary speculations to his 1909-1911 magnum opus, specifically linking interneurons with the generation of the limited reality we experience: "*During its life, each animal therefore does not have, in its brain, all the possible representations, nor all the possible associations of these. It has only the representations and combinations of representations currently most useful for the conservation of its species. The spirit of severe economy which reigns in nature allows it, in fact, to have only the cerebral apparatus of association most suitable for its defence and the perpetuation of its species, in a word, the apparatus best suited to its current conditions of existence. But come new conditions, new needs, like those created in humans by civilization, and most probably the mechanism of these brain neurons of association is changing and improving. It may also be that, under the influence of these new conditions, aberrant intercellular relations occur, which would explain well the genesis of certain secular errors with deep roots; such as the belief in free will, among others.*" (Cajal, 1909-1911).

In addition to the classic butterfly-like interneurons, Cajal described a fourth type of neuron in "higher vertebrates" such as human, the psychomotor neuron, which is also between two neurons. The psychomotor neuron is now commonly known as a projection neuron. Here are Cajal's comments: "*the psychomotor neuron picks up, with increasing clarity, a larger set of more complicated impressions; from there, it transmits, with ever-increasing precision, its orders to the other nervous centres, developing here the activity of the motor neurons, inhibiting there the automatic reactions, the reflexes arising from the ganglia. He thus centralizes, in him, psychomotor corpuscle, the functional solidarity of the whole being, solidarity just now rough, despite the neuron of association. Then, without a doubt, awareness of personality and unity, memory, intelligence, will are awakened, much more sophisticated means of defence and conservation. The being entirely dominated by the psychomotor neuron is more and more powerful in the struggle for life.*" (Cajal, 1909-1911). What Cajal inferred over a hundred years ago is in line with the process of sensorimotor transformation.

"*The soul never thinks without a mental picture.*" Aristotle (384-322 BC)

In 1930, Edward Chace Tolman and Charles H Honzik showed that rats actively build an internal visual image of their external environment, which allows them navigating complex mazes (Tolman and Honzik, 1930). Tolman referred to this representation of the features and landmark of the external environment as a "cognitive map". He postulated that, with multiple exposures to a maze, rats memorized multiple environmental cues to build increasingly more accurate internal representations of the maze. As Cajal, Tolman was also a visionary. Based on his findings, he hypothesized that humans do not passively respond to their external environment but rather act on beliefs, attitudes, changing conditions, and that they are driven by goals. In the seventh chapter, we will explore how these cognitive maps are associated with various cognitive biases.

At a more encompassing level than cognitive maps are the actual mental models, which our brain manipulates not only to represent our external and internal environments, but also to simulate diverse outcomes and chose the most adaptive output. The profoundly impacting concept that we unconsciously create mental models as an interface between sensory input and motor output was first proposed by Kenneth Craik, in his 1943 book "The Nature of Explanation".

Craik put forward that thinking consists of manipulating internal representations of the physical world, which we have evolved to interact with. Here we have our second protagonist. During his short life (March 29th 1914 - May 7th 1945) Kenneth Craik became a pioneer of cognitive science by articulating that the mind is a symbol processor, exhibiting the unique ability of making predictions used for generating outputs. Whether Cajal ever met Craik, let alone was exposed to his ground-breaking ideas, is unlikely. Fantasising on an imaginary gathering of the two minds (figure 25), what would they have said to each other?

In 1934, Craik was 20 years old when Cajal died at the age of 83. Had they merged the concepts they each pioneered, both thinkers could have concluded that "thinking" and the "sensorimotor transformation" are one. Craik defined three steps involved in thinking:

Figure 25

1) Translating specific external processes (sensory inputs) into an internal representation in terms of words, numbers, or other symbols.

2) Deriving other symbols from these sensory inputs by some sort of inferential process.

3) Re-translating these symbols into an output, which can be motor or awareness that a prediction has been achieved. In this manner, thinking allows considering possible futures, to ultimately generate the most adaptive output.

The foundations laid by Craik established that thoughts and feelings are indissociable; they represent the same symbol-based predictive process. Craik considered that the brain had evolved to be a predictor, while what we call the mind or soul are consequential to this evolutionary process. His words:

"My hypothesis then is that thought models, or parallels, reality — that its essential feature is not 'the mind', 'the self', 'sense-data' nor propositions but symbolism, and that this symbolism is largely of the same kind as that which is familiar to us in mechanical devices which aid thought and calculation."

"If the organism carries a 'small-scale model' of external reality and of its own possible actions within its head, it is able to try out various alternatives, conclude which is the best of them, react to future situations before they arise, utilize the knowledge of past events in dealing with the present and future, and in every way to react in a much fuller, safer, and more competent manner to the emergencies which face it."

By including Craik in our reflexion on mental models we add the element of "prediction". I program what I am going to do so that my action is optimally adaptive with the most likely prediction. Think of the Hering illusion presented in the first chapter (figure 18). What we experience in this visual illusion is the prediction of what will happen in about 300 milliseconds. When a projectile, such as Bertrand Russell's rock, is directed at you, a rapid action is elemental to avoid a potentially fatal outcome. If we could go back in time, we could witness a most dramatic example involving our second protagonist. It was May 7th 1945, two days prior to the end of the Second World War in Europe. Kenneth Craik was overworked: heading covert research for the war efforts during the day and doing his own personal research and writing at night. Craik was riding his bicycle in central Cambridge. O.L. Zangwill recounts in his 1980 homage to Craik (Zangwill, 1980): "*Never an accomplished cyclist, he came into collision with the partly open door of a stationary car in King's Parade and was thrown from his machine and apparently struck by an approaching lorry. He never regained consciousness and died later the same night. At the inquest, a verdict of accidental death was returned and no blame was attached to either driver.*" A moment of inattention abruptly terminated Craik's prescient quests.

While he remains largely unknown – Kenneth Craik did not even make the list of the one hundred most influential psychologists (Haggbloom, 2002) – there is still so much to learn from Craik's acumens. The philosopher of mind Dan Williams mentions; "*I think that Craik had in mind a wide range of potential uses of prediction: The capacity of the brain to predict the likely sensory consequences of the organism's behaviour in order to overcome signalling delays. Flexible predictive tracking mechanisms. The ability to use predictive models to represent features of the world the organism does not currently have access to*". (Williams, 2018).

"I am blinded. I open my eyes wide and only see. But the secret – that I neither see nor feel. Could I be making here a true orgy of what's behind thought?" Clarice Lispector (Água Viva)

The reality we experience is not only a model but also a time predictor. The statement that our brain predicts potential futures would still surprise most. The time window of the "future" depends on the nature of the events that dictate the required adaptative output, i.e. survival and reproduction. On the small scale, we have the Hering illusion pointing to a prediction in the range of 300 milliseconds. There is another example. Think of the cocktail party, and your line that we are blind about 4.5 hours per day. Yes, microsaccades, the unconscious movements between fixations, during which you are blind (4 hours), and the blinking (30 minutes). These micro movements keep the eyes steady in expectation of visual stimuli. There is extensive experimental evidence that microsaccades are suppressed just before a visual event is predicted (Amit et al., 2019). It makes common sense; if a time window of 300 milliseconds can determine life or death, then one cannot afford to be blind during this decisive moment. Therefore, when the brain predicts a potential event, the microsaccades stop. Seeing a threat – visually – is associated with stopping microsaccades within the time prediction of danger to one's physical integrity: a tiger jumping at you.

However, more intriguingly is that these microsaccade inhibitions are not only restricted to visual predictions. It is perhaps counter intuitive, but the prediction of touching something also coincides with the inhibition of ocular microsaccades.

Here is the experiment: human subjects are asked to perform a purely tactile task, that is, finding tactile cues with their fingers. No vision involved, just touch. Imagine the participants fixating a point on a computer screen in front of them, while both hands are laid on a table, either side of the computer. One hand was used to detect the tactile stimulus while the other was used to press a button. While fixating a computer screen in front of them, the task was to press a button with one hand when they felt vibrations to the tip of one finger of the other hand, being faster or slower than 60 Hz. When they expected a change in vibration frequency, the subjects stopped microsaccades. One might ask what is the point? Does microsaccade inhibition play a role in finding a tactile target with your finger? Yes it does. The investigators showed that in instances when microsaccades occurred around the time of tactile target presentation, the performance was reduced (Badde et al., 2020). The gist here is that our brain performs intricate tasks with one ultimate goal: predicting the future to produce the most adaptative output.

These tasks entail a crosstalk between sensory modalities. Here we have an example of vision and touch, but this is likely the tip of the iceberg. There is so much more to learn about how our brain generates reality and predicts the most likely futures and coordinates the most optimal output. Time scales encompass a wide range.

Now on to a longer scale, here is a famous and well-accepted example with a different context: Pavlov's dogs were conditioned to have access to food only after hearing a metronome. Once trained, the sound of the metronome alone induced salivation. While time scales and context vary extensively in the process of "thinking", the commonality is that, ultimately, species have evolved to produce optimal predictive adaptative responses for survival and reproduction.

Prediction is essential for adaptive success. As Craik put it: *"one of the most fundamental properties of thought is its power of predicting events,"* (Craik, 1943). He added: *"only this internal model of reality — this working model — enables us to predict events which have not yet occurred in the physical world, a process which saves time, expense, and even life."* How so very ironic that Craik's death involved the convergence of external events and the lack or perhaps the impossibility of an adaptative response, leading to a fatal outcome. At least two *Homo sapiens* and their respective interactions with technological apparatuses (the one who opened the car door and the lorry driver approaching fast and perhaps too close to the stopped car) contributed to Craik's premature death.

What did Craik mean by "adaptative success"? *"Living organisms are in a state of dynamic equilibrium with their environments. If they do not maintain this equilibrium they die; if they do maintain it they show a degree of spontaneity, variability, and purposiveness of response unknown in the non-living world. This is what is meant by 'adaptation to environment' ... [Its] essential feature... is stability — that is, the ability to withstand disturbances."* (Craik 1966, The Nature of Psychology, pp. 12-13). Craik put forward that the predictive nature of sensorimotor transformations is flexible, which went beyond the simple correlations between sensory inputs and outputs described by behaviourists in his days (Williams, 2018).

Cortical and subcortical circuits create internal models as a program for action. To interact with the physical world, sensory inputs update such models (Huda et al., 2018). As we mentioned before in the first chapter, sensory information modulates the construct of reality by our brain. Our eyes are not cameras sending images to the brain; rather, our eyes modulate the images created by the brain. Close your eyes and you will "see" images.

Moreover, sensory information itself is influenced by previous experience, adding another level of subjectivity. The decisions about what we are doing depend on unconscious internal models that are constantly built and updated by learned sensory inputs, a process known as "evidence accumulation". Studies of neuronal activity in animal models have shed light on the actual underlying mechanisms. Sticking with vision as a sensory modality, we can pinpoint at least two brain regions, both in the cortex: 1) the lateral intraparietal area of the posterior parietal cortex; and 2) the frontal eye field division of the prefrontal cortex (Kim et al., 1999; Shadlen and Newsome, 2001).

The neural mechanisms that dictate what we perceive (sensory) and what we do (effector) are inter-dependant. There are dynamic inter-relations between input and output. "*Perception and action are co-determined, and sensory inputs that lead to perception derive meaning in light of task- dependent goals and the actions that are selected to achieve them*" (Huda et al., 2018). One might think such feature is unique to us humans, generously including primates, perhaps. Most findings were obtained using rodents.

Recent findings from Sten Grillner revealed that lampreys, which date back to more than 560 millions years, possess internal maps of the outside world (retinotopic and somatotopic), respectively built from visual and tactile sensory inputs conveyed through the thalamus (Suryanarayana et al., 2020). Why does this matter? We erroneously ignored animals that lacked a cortex and thought that these older ancestors essentially relied on olfactory inputs for their survival and reproduction. We were wrong. Lampreys have the equivalent of a cortex, known as the pallium.

We also ignored birds for similar reasons, despite the visionary work of Harvey Karten published more than half a century ago, which clearly established the occurrence of similar visual information processing as in human, such as binocular vision (Karten, 1969).

Harvey Karten correctly pinpointed the dorsal ventricular ridge and adjoining dorsomedial pallial 'wulst' as the regions capable of refined sensory-motor and cognitive operations.

There is accumulating evidence based on Karten's findings that the brains of birds possess all of the features allowing the generation of a multisensorial reality. Moreover, there is also evidence from drosophila (fruit flies) of simpler, yet defined, mental models called associative loop (Fiore et al., 2015). Figure 26 shows basic elements of mental models in primates (A), lampreys (B) and drosophila (C).

Figure 26

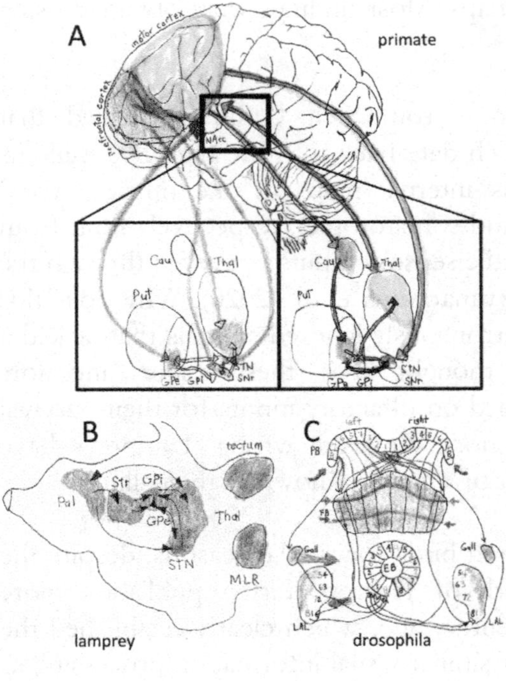

Back to our first protagonist, Cajal: he also foresaw that brains are characterized by their ability to generate a single and integrated perception that corresponds to all the stimuli, which allows interacting with the external world (chapter 48 of Textura del sistema nervioso, 1899).

Taking inspiration from the father of experimental psychology, William James, Cajal also put forward that these multisensory representations are "plastic", alongside the neurons and networks they form to generate these representations (Ferreira et al., 2014). Even in the fully developed brain, neurons can move and change shape. Considering that the healthy adult human brain looses over a thousand neurons every day, such level of plasticity is essential. Your neurons die all the time. Neuronal death is a fact of life. The synapses that neurons establish between themselves are also remarkably malleable.

Simply remember the change in retina circuitry when transiting between different background light levels. As you wake up, transiting from dark to light, the levels of dopamine and nitric oxide increase across the retina, proportionally. Exposure to dim light prevents electrical coupling between rod bipolar cells and AII amacrine cells while permitting electrical coupling between rods and cones. Under daylight, coupling ceases between rod and cone photoreceptors. We can regard these events as a switch preventing any rod signals from modulating brain activity, favouring fine detailed vision mediated by cones. Just imagine if the highly sensitive rods, with their poor visual acuity, were to contaminate our daylight vision? The changes in retinal circuitry also contribute to maintain contrast sensitivity relatively stable over a colossal range of light adaptation covering 10 to the power of 10 units of light levels.

Example of plasticity: Sensory substitution

Technology is at a level that electronic devices can substitute one sensory modality for another. The application is highly relevant in blindness. A blind person can receive mild pin-like stimulations across their back driven by a camera, above their head, which transduces visual signals into touch. Even subtler is when the touch stimulations are distributed across the tongue, and a miniature camera and transducer are discretely attached to special glasses. Another ingenious approach is to transduce pixels from a camera input into sequential sounds of different pitches. With training, sounds elicit activity in the visual cortex. These are powerful examples of the brain's capacity for plasticity (Proulx et al., 2014).

History puts our second protagonist devoting his genius to tackling the puzzle of plasticity (Craik, 1938, 1941). Building his own tools, Craik focused on the mechanisms underlying light adaptation as part of his doctoral dissertation, published in 1940. That was five years before his untimely death at age 31. He applied his findings to improve night vision in airplane pilots during the Second World War. In retrospect, Craik was onto something highly relevant: how does the brain build the reality we experience?

One cannot help but ponder the wonders that Craik's beautiful mind might have produced had he lived longer. What links would he have made between neural circuits and behaviour? Let us allow the fantasy of conversing with Craik about retina-based light adaptation, and more, more of the many instances in which diffusion of molecules can modify neural circuits. Such "wireless" effects typify how the same neuronal circuit can mediate drastically different behavioural outcomes (Jékely et al., 2018). We see this happening in the limbic system, underlying emotional states, where molecules known as neuropeptides act at a distance inside the brain to control synaptic communication and dictate how we respond to internal and external stimuli.

So many subjects could have fuelled Craik's astute conception of predictive mental models. Indulge me in one more example, which I think you will find exceedingly pertinent with regard to the plasticity of how our brain constructs representations of our body.

I would have asked Craik to participate in the rubber hand illusion. He would have placed both hands on a table, but a partition would have prevented him from seeing his left hand. Instead he would have seen in front of him, a rubber hand, which of course he would have known was not his own hand. Here is the trick: I would have simultaneously brushed both his left hand and the rubber hand with a paintbrush, synchronizing changes in pressure, rhythms and locations on his different fingers. After ten minutes, Craik might have "appropriated" the rubber hand in front of him as his own left hand. Feel free to watch videos by searching "rubber hand illusion". Finally, I would have mentioned how the neurologist Vilayanur Subramanian Ramachandran developed an approach to "move" phantom limbs.

Patients who have undergone amputation of a limb can experience the sensation of this "absent limb". In some cases the sensation is painful because of its felt position and stillness — yes we are talking about mental models of non-existent limbs. Remember, your brain generates what you feel, regardless of whether it is real or not. Using mirrors, Ramachandran makes the patient see the opposite limb on the side that corresponds to the amputated limb. After some training, the patient appropriates the intact limb as the amputated limb, allowing finally the phantom limb to be moved in a more comfortable position (Steenwinkel et al., 2019).

With this information, Craik would have offered links and context on how the brain is wired to generate the reality we experience and to provide unconscious predictions that determine our actions. Just imagine for one moment what Craik would have accomplished if he had had access to functional brain imaging technology.

For the grand finale, with the luxury of Craik living still, he would have witnessed what virtual reality combined with head mounted cameras showing one's own body and multisensory stimulations can do: nothing less than "out of body experiences". This ultimate illusion is known as an "autoscopic experience" (Aymerich-Franch et al., 2016); from ancient Greek αὐτός ("self") and σκοπός ("watcher"). I bet you Craik would have loved to be a subject. He did experiment on himself a great deal, to an excessive and dangerous extent. It is reported that Craik looked directly at the sun and poked his own eyes (Zangwill, 1980). Over more than 75 years after his death, it feels as if we just barely have begun percolating Craik's profound insights. But let us be optimistic, the data are coming (even from magicians, or should I say: illusionists), we need to digest them, to make them available, because this is reality. Of note is that there have been concrete applications in artificial intelligence and robotics. Artificial neural-networks are based on Bayesian models. This means that the robot's artificial neural networks must be able to infer the cause of the sensory inputs, as opposed to the sensory inputs themselves modulating the network, this is called a top down model.

The notion of top down means that each of the robot's perceptual worlds depends on its own individual imaginative abilities, on its computing power to calculate the respective statistical probability of each imagined worlds. Because there are almost limitless imagined worlds, Bayesian-based sensing cannot be completely flawless. Robots need to be trained by previously exposing them to mindboggling permutations of possible environments in order to recognize the nature of specific sensory inputs.

Could computer models "see" optical illusions? The answer is yes: they can be fooled, as we can. In line with a Bayesian approach, computerized neural networks must already have a programmed model (algorithm) of each of the distinct visual aspects that together create the illusions and be trained by repeated exposures to the illusions (Mély et al., 2018). It is highly unlikely that our own brain works in a strict Bayesian fashion (Williams, 2018). However, a Bayesian approach does provide promising results with artificial intelligence, such as in designing machines that learn from the data acquired through experience (Ghahramani, 2015). In chapter VIII, we do discuss the other side of the medal: artificial intelligence can be a wild beast to tame.

The science of magic

"... *magicians hack the main feature underlying our intelligence and construct a biased interpretation in the spectator by using different forms of misdirection. The surprising resolution of a magic trick is, in this sense, comparable to the effect that can be produced by a visual illusion or by being suddenly aware of a false memory. In all these cases, unconscious assumptions have been broken.* (Quiroga, 2016)

Magic is a powerful tool to decipher how mental models are built. Functional brain imaging of subjects being misdirected when viewing magic tricks had heightened activity in two regions: the left dorsolateral prefrontal cortex, involved in monitoring cognitive conflicts, and the anterior cingulate cortex, involved in resolving these conflicts (Parris et al., 2009).

There is a violation of what logical outcome should result from a familiar action. Functional brain imaging of subjects experiencing "impossible outcomes" showed higher activity in a region called "the head of the caudate nucleus". This region is involved in processing changes in the contingency between action and outcome (Danek et al., 2015).

In his book, "Experiencing the impossible: The Science of Magic", Gustav Kuhn explains how magicians are astute exploiters of how our brain generates the reality we experience. They demonstrate how the brain constructs a subjective reality using assumptions based on relatively little and ambiguous information.

Figure 27.

Back to the human brain, there are two important aspects for predicting potential futures and for generating the most adaptative response: learning and memory. In fact, from an evolutionary perspective, learning and memory evolved to do just that. Pavlov's dogs are a flagrant example: they associated that a reward (food) becomes available only after the presentation of a specific sound: that of a metronome. The process of learning and memory accounts for their salivation when exposed to the metronome sound alone.

Another example, from a family life, involving a carton of milk: as a caring single parent, home alone in the evening you ventured in the rain to buy milk from the corner store, planning for tomorrow's family breakfast. You get up early, earlier than the whole family; half asleep, you open the fridge door, reaching eyes half-opened for the milk. On automatic pilot, you aim precisely at its location without even thinking. Also flawlessly, you grasp, and you lift, but unexpectedly you are brutally pulled out of your lethargy by the rocketed projection of the milk container up in the air... you unconsciously over-estimated its weight. Your learning and memory prepared you for a heavy receptacle, but not for the nocturnal indulgence of your three teenagers in copious cereal bowls, alongside clumsy spills. At this moment, your senses capture the dubious puddles on the floor by the empty bowls laden counter. You also remember the late night escapade your teens had and you, now fully awake, deftly reconstruct the scene of the crime.

There is accumulating evidence linking neural networks and behaviour. Learning and memory also entail modifying neural networks both anatomically and biochemically. What if I told you that this plasticity and the associated behaviour could be transferred between two organisms? Bédécarrats et al. (2018) trained sea slugs (Aplysia) to learn a specific task, and then transferred genetic information in the form of RNA to untrained Aplysia, which provided them with the ability to remember. Without training, naïve Aplysia learned solely based on having received genetic information and translating it into proteins. Which proteins you might ponder? In neurons that mediate several forms of long-term memory in Aplysia, the DNA repair enzyme poly ADP ribose polymerase 1 (PARP-1) is activated. In virtually all of the eukaryotic cells tested, the addition of polyADP-ribosyl groups to proteins (polyADP-ribosylation) occurs as a response to DNA damage. Thus the finding of activation of PARP-1 during learning and its requirement for long-term memory might come as a surprise. Cohen-Aromon et al. (2014) suggested that fast and transient decondensation of chromatin structure by polyADP-ribosylation enables the transcription needed to form long-term memory without strand-breaks in DNA.

The requirement for polyADP-ribosylation to long-term memory formation also extends to mammals such as mice (Goldberg et al 2009). The conclusion is that information about learning and plasticity can be passed on between species by modulating the expression of specific genes. In an experimental setting, we can modify a determined behaviour by transferring genetic information (in the form of

RNA), at least in gastropods and potentially in mice; the field is still in its infancy.

Another mean of modulating the expression of specific genes is epigenetics. Strictly speaking, epigenetics refers to an inheritable ability for certain genes to have their activity modified without needing to alter their sequence. Changes in activity typically involve DNA methylation and histone modification through non-genetic factors such as external agents or as part of normal development. The list of agents is almost limitless: it includes your diet, where you live, your social interactions, your education, your sleep pattern, your physical activity, and diseases such as cancer, diabetes, Parkinson and Alzheimer's, just to name a few.

Factors susceptible of being associated with turning on or off the expression of specific genes (and their specific function) encompass basically any aspects of how you were brought up starting from the womb all the way to aging. Susceptibility to these factors is what makes us all different. We all have minor differences in gene sequences, which are not mutations; they are called "polymorphisms". These polymorphisms are passed along generations; hence the inheritable nature of epigenetics. Care must be taken to extract signal from noise in this jungle where again one can attribute changes in neural networks and behaviour by solely invoking epigenetics.

It would be counterproductive here to navigate the intricate meanders of this vast expanding field, which touches all aspects of biology, from evolution, to development, cancer and most diseases, including mental health. For instance, there are numerous studies reporting that polymorphisms in genes involved with serotonin neurotransmission are associated with depression and anxiety (Gottschalk et al., 2017). At the moment, results are often conflicting and some attempts at replicating studies have failed. Regardless, the field represents a new frontier for research and will occupy us for many years to come. What is being revealed here is that our environment does affect how our brain is wired, and therefore how we might perceive and anticipate reality.

If we recapitulate, our brain generates the reality we experience and our brain also predicts the most likely events in order to generate optimally adaptative responses. By optimally adaptative, I mean the responses that will promote survival and reproduction. These are the two reasons why we have evolved, why DNA has been passed on over 3.8 billions years across generations (vertically) and between organisms (horizontally). Our brain has evolved to detect only the physical modalities that relate to survival and reproduction. Our brain detects these features (internal and external to our bodies) by modelling these very physical aspects. As mentioned above, as an offshoot of this capacity to model the world using sensory inputs, our brain produces models constantly, which are most often not fed by sensory modulations. Learning and memory feed these models and predictors. Moreover, what we just flew over with epigenetics is that these environmental factors themselves can determine how our brain generates and anticipates reality. By producing models without sensory input and motor outputs, the brain can act as a hamster in a wheel, endlessly spinning, ruminating. This exaptation can be a curse or could be a superb source of infinite creativity. The distinction between both is blurry, highly imaginative people often walk on a fine thread. The curse comes when spontaneous models represent false threats fed by boundless inventiveness.

One notable aspect as threat detectors is a region of the cortex called the amygdala, which receives sensory inputs — directly from the sensory thalamus for vision and hearing or from the various sensory regions of the cortex for smell, taste and touch. The amygdala acts as a gate that determines whether a distress signal should be sent to the prefrontal cortex so that we become aware and can act accordingly. A simplistic view is that the neurotransmitter serotonin prevents the signal to be sent. Lower levels of serotonin release are associated with heighted levels of anxiety and depression. The prefrontal cortex receives "false threats". Persistent signaling of false threats is maladaptive, but can be threated. Some antidepressants act by blocking the reuptake of serotonin at the level of these synapses, which prolongs the action of serotonin and diminishes the threats signals. These drugs are known as SSRIs (selective serotonin reuptake inhibitors). The point is, one can make a link with the strong interest of the epigenetics of genes involved in serotonin signalling as a mean to better understand why some people are more prone to suffer from anxiety and depression. The truth is most of us do experience anxiety and depression at some point in our lives, but at different levels, duration and frequency.

One important point related to mental health is that the reality generated by the brain goes beyond the physical world itself; it generates your sense of self. But we are oblivious to all these processes. "*Because we are our brains, which create our sense of self, we have no privileged access to this invisible process from an outsider's perspective*" (Bruce Hood, The Self Illusion). We are oblivious to the mechanisms that create what we consciously experience; this is true for all sensory modalities as well as for the sense of self. In fact, your own sense of self is a creation of your brain, based in part on how you think your social world (real and now increasingly more so digital) perceives you. Bruce Hood puts forward that it is the existence of others and our interaction with them, direct and indirect, that defines who we are. Another quote from Bruce Hood's "The Self Illusion": "*We cannot be aware of the underlying processes that create it and yet we feel it as coherent. You never see that you are a reflection of the others around you because you cannot easily see how you change. And we don't easily see our self switching from one to another until it is pointed out to us by those around us who recognize that the context has changed*". Examples of changes include: changes in accents, mood, being more charming or defensive, or overcompensating due to a feeling of insecurity or low confidence.

We have a tendency to linger in the past. Our brain is wired to ruminate past memories. Together with this apparent curse, our brain protects us from the true nature of the external world and from the mechanisms taking place to create what we perceive and experience, such as what we "see" and "feel we are" (self). "*Cognitive dissonance protects us from ruminating over failed goals, positive biases keep us motivated, free will gives us grounds for praise and blame (and ownership of our actions, conducive to morality), decision making gives us the illusion of control. Without these cognitive illusions, we would not be able to function because we would be overwhelmed by the true complexity of the hidden processes and mechanisms that control us*" (Bruce Hood, The Self Illusion).

"*I thought these grapes were ripe, but I see now they are quite sour.*" Aesop (Aesop's Fables)

Here lies a paradox, one that is present in this very book: we are drawing our attention on that which has evolved to be hidden. However, we are not trying to be conscious as the mechanisms happen, but our endeavour is to learn about their nature, to be humbled, to be empowered, to have a more encompassing "view", "grasp", and awareness" of the reality we navigate in. We are trying to become more mindful. Yes, mindfulness is a trendy word... mindfulness... just being able to stop and observe. In her book "When Things Fall Apart" Tibetan Buddhist Pema Chödrön states: "*We have a concept of ourselves that we reconstruct moment by moment and reflexively try to protect.*" As threat detectors, our brain has evolved to protect against all threats, real or perceived. Reflecting on these false threats persistently manifesting themselves as revving thoughts, Chödrön proposes that we "... *get to know them, see how they hook us, see how they colour our perception of reality, see how they aren't all that solid.*" Adding my ten cents to her wisdom, I would say: observe your perceived pain realizing there is no threats, no threats, live your pain for the creation that it is... wanting to change this feeling will give it a life of its own that it does not have.

Back to the text with numbers and letters in the prologue, if you do read it mindfully, you will take in overlapping layers, a message and noise. We extract meaning in noise. Being mindful means simply recognizing and acknowledging biases. However, we are rarely mindful, unless you devote your life to meditation, contemplation, mindfulness, or stillness. Even then, your experience of reality generated by your brain remains biased, and personal. As Steven Pinker puts it in his book "The Stuff of Thought":

"Humans construct an understanding of the world that is very different from the analogue flow of sensation the world presents to them. They package their experience into objects and events. They assemble these objects and events into propositions, which they take to be characterizations of real and possible worlds. The characterizations are highly schematic: they pick out some aspects of a situation and ignore others, allowing the same situation to be constructed in multiple ways."

We are connection detectors, pattern detectors, we can see a pattern in random events and call it ghost, spirit, paranormal... we ignore all the noise and fixate on what might be a pattern, like the apparition of a face in random noise... yes, especially faces. For example: 1) The Old Man of the Mountain in Franconia, New Hampshire, USA; 2) Roche Bonhomme in Jasper National Park, Alberta, Canada; and 3) "Smiley face" in Galle Crater, and "Face on Mars" in the Cydonia region, both on Mars, which was highly unlikely visited by aliens, at least not as we anthropomorphize them. The manner in which we detect patterns in randomness (a phenomenon called pareidolia) is in fact pretty telling of our personality. The "Rorschach inkblot test" banks on pareidolia as a mean to obtain information about a person's mental state. Study subjects are asked to verbalize what they "see" in standardized random inkblot patterns.

Another perspective to appreciate our biases is the argument by Alfred Korzybski that we will never be able to fully understand the world because of the limitations imposed by our nervous system and the languages we developed. In his book "The Map is Not the Territory", Korzybski articulates that brain and language thus represent filters that preclude an objective comprehension of reality.

Back to the carton of milk, which led to an overestimate of its weight and a spill. Bloody teens, but we still love them. Learning and memory prepared you for a heavy receptacle, but not for the nocturnal indulgence of your three teenagers in copious cereal bowls, leaving the carton of milk almost empty. Lets expand on this example as a mean to understand the highly personal way in which we construct reality, continuing with Steven Pinker's "The Stuff of Thoughts". We rely on an inventory of basic units: events (I bought the milk late last night, I woke up to prepare breakfast); states (I was not quite awake when I walked to the fridge); things (the fridge, the door, the shelves, the carton of milk); substances (a liquid, milk, filling a square carton); places (the position of the milk in the fridge); and goals (getting the milk out of the fridge).

Your brain analyses (in distinct regions) all these units to produce a unified action that explains why you successfully grabbed the milk but exerted too much power in pulling it out of the fridge. I am oversimplifying, because your whole life's background does come into play, perhaps in more complex situations, but still here as well. Your action of grabbing the milk carton was planned without paying too much attention, which reinforces that the brain has an intrinsic ability to analyse independent units and approximate a reality, which you can navigate, without having to consciously think about all these units. Here we saw a mistake, because pertinent units were not accounted for in the analysis: the teens drinking the milk in the middle of the night. Analogously, the visual illusions we talked about in the first chapter illustrate how flaws can happen in the process of your brain decomposing the visual scene in various properties and binding them together to produce a unified visual experience.

"*Reality is merely an illusion, albeit a very persistent one.*" Albert Einstein

The ultimate evolutionary drive for the generation of a reality is the detection of sources of threats, pro-survival, and pro-reproduction. Detection of threat and pro-survival aspects of our environment are typically associated with pain and reward-pleasure, respectively. Furthermore, the construct of reality corresponds to: 1) past events for learning and memory and 2) predicted events (think of the Herring illusion). It is hard to live in the now!

When we feel bad in the moment, we think our current states represents our future. Our perception is so far from the truth. We'd like our state to change now, now, now. When we feel bad, we can repeat this sentence again and again: **my current situation is not my permanent destination**.

"*We all tend to assess the future merits of any activity based on what it offers at the moment.*" Bruno Bettelheim (Psychanalyse des Comptes de Fées, 1976)

"*A human being is a part of the whole called by us universe, a part limited in time and space. He experiences himself, his thoughts and feeling as something separated from the rest, a kind of optical delusion of his consciousness. This delusion is a kind of prison for us, restricting us to our personal desires and to affection for a few persons nearest to us. Our task must be to free ourselves from this prison by widening our circle of compassion to embrace all living creatures and the whole of nature in its beauty.*" Albert Einstein

Chapter V

The Meaning of Life

You ask, what is the meaning of life? You do so because you can. Your inclination and ability to ask this question are incidental to the evolution of predictive mental models. These models emulate and predict the specific physical aspects of reality that have determined survival and passage of genetic information from and between species over 3.8 billion of years.

The fact that you can ask this question is an offshoot of your brain's ability to model and predict a reality without requiring sensory input or motor output. Your brain can now ruminate freely and endlessly. The urge we have to ask, "what is the meaning of life", is linked with our nature of being threat detectors. We have evolved because we detect threats and respond to them, all species do. The lack of sensory awareness to the neuronal activity that generates predictive mental models, creates a void, and becomes a threat in itself. Where are these endless thoughts coming from?

You model a world and fear threats without any sensory inputs and motor outputs. This can be a curse. Why do we live, why do we die, who are we, what is the point of all this? The moment in evolution when predictive mental models became independent of sensory inputs and motor outputs, existential crises were born.

You will say: this is too complicated an answer and does not really satisfy me. You need an answer. What is the meaning of life? This is a question you likely asked yourself at some point and perhaps still do, since you did grab this book, curious about its very title. Ha-ha, I got your attention! You found complicated figures, text with scientific jargons; surely the meaning of life does not entail knowing all of this. If you did not jump straight to this chapter, you might have suffered through all this dry material. You might not even be reading this, having given up on the density or seemingly scattered focus of this book.

I, the author, say that philosophy is a private club for members only; it can be so opaque. Well, so can be science. But the scientific method, at its very foundation, is the best tool we have "to observe", yes to observe the reality around us. Some scientists and science communicators are more apt at making it accessible. With the shortening of attention span fostered by electronic media, educators and communicators need to have pizazz; some have whole teams devoted to their endeavours. If you read the reader's right, perhaps it brought you directly here, to chapter V. Then why not cut to the chase and just write a book based on this chapter, you might also ask? Why all this science, often dry and inaccessible to many?

Context, ones needs context. If you did jump straight to this chapter, please allow me a recap for context's sake. I promise, lay language, compassionate to all types of readers.

The first chapter makes the point that the brain constructs the reality you experience and can do so even without sensory stimuli. I used vision as an example since this is my field of expertise. The eyes do not act as cameras; if they did you certainly would not experience visual images when closing your eyes or when dreaming, and hallucinations would be hard to explain. Also, we fall for illusions because they result in discordances between the images observed and how the brain constructs them. Starting in the eyes, images are decomposed into specific aspects that are sent to the brain to be analysed separately in distinct regions, to ultimately produce a unified coherent visual experience.

The second chapter makes the point that our senses – and how we react with our internal and external environment to survive and reproduce – have evolved from single cells dating 3.8 billion years. By following the evolution of these single cells, we can track similar features for sensing and reacting. Allow me only one word with a science flavour: transduction. Transduction means changing one form of energy into another. Taking vision as an example, light energy is transformed into a chemical reaction. The manner in which mostly all known species sense light starts with the same molecule in single cells dating billion of years. When hit by light, this molecule changes its shape, which triggers a series of chemical and electrical events culminating to the creation of mental images (conscious or not) and an output (voluntary or not) if required for survival or reproduction. Another point is that most of the evolution of sensing took place in single cells, which were the only form of life on earth for the first 3.258 billion years. Picture this stretch of time: ample time for trial and error, with genetic mutations and exchanges of genetic information between cells.

The one common variable in life is the exchange of genetic information. As I mentioned in the introduction, we carry genes… once we accept it, we are free to live; we can understand why we ask: "what is the meaning of life"? We'll get back to that. The last 0.542 billion years, which represent only 14% of life's timeline on earth, saw the evolution of all multicellular organisms; yes all of them. Grouped as one organism, cells became specialized. In a figurative way, single cells sacrificed their own survival by being part of multicellular organisms. Some cells became sensors, while others became effectors. Between these cells, interneurons evolved to generate a representation of the environment in which an organism survives and reproduces.

The take-home message here is that the process of evolution is continuous. No external agents came into play at some point so that novel species became conscious or suddenly had a soul. If you are searching for site of the soul, or mind, or if you prefer "thoughts" — and you are wondering why we have the ability to ask the question "what is the meaning of life" — then you ought to be looking between sensors and effectors.

The third chapter presents the various levels linking sensors and effectors. Single cells rely on the first level since they possess both the sensors and the effectors. However, do not think the linkage between both is simple. Roaming the earth alone for 3.258 billion of years, single cells evolved highly intricate manners of sensing and acting with their internal and external environments. Numerous genes with highly specialized functions evolved during this very long period of time, allowing for countless mutations and gene exchanges and a myriad of trials and errors to determine which mutations were adaptative or not. We also find this first level in the guts and lungs, where the same cell can "feel" changes in its environment and "act" by participating in the movement of food and the clearance of mucus, respectively.

The next levels apply to multicellular organisms. At the second level, sensory cells connect directly to an effector cell, such as a cell possessing cilia for underwater propulsion.

At the third level, a sensory cell relies on a separate intermediate cell (motoneuron) to connect with an effector cell, which is typically a muscle cell. A well-known example is the knee-jerk reflex, where a muscle stretch sensor sends a signal directly to a motoneuron in the spinal cord, which in turns tenses a muscle. However, this is not sufficient to spontaneously raise your leg. Another muscle must be relaxed at the exact same time. This requires the fourth level. To silence the other muscle, stretch sensors also send a signal to interneurons, which invert the signal (via an inhibitory action), therefore silencing the motoneurons, which in turn relaxes this other muscle, producing the knee-jerk reflex.

You guessed it; the fourth level is the domain of the infamous interneurons; the cells that interpose themselves between sensors and effectors, forming networks that can coordinate many sensory cells and effector cells. Examples include all that our body accomplishes automatically, such as breathing, walking, chewing, and bowel movement; one refers to the part of the brain involved here as the "reptilian brain".

The fifth level refers to when these complex networks of interneurons start to "have a mind of their own", that is, when they can model internal and external environments without relying on sensors or effectors. This last level is discussed in the fourth chapter. But before we talk about the fourth chapter, let us wrap up at this point. What we saw is that as evolution followed its course over 3.8 billion years, sensors and effectors became separated by intricate neural networks. Eventually, assemblies of neurons formed a predictive model of the reality in which species had to survive and reproduce.

In chapter IV, we related how in the early 1900s, Ramon y Cajal was able to visualize and characterize in extensive details the cells that interpose themselves between sensors and effectors. We then introduced Kenneth Craik who proposed that our brain generates mental models as an interface between sensory input and motor output. Craik also put forward that such mental models served as predictors of the likely events to happen in the immediate future. In this manner, the most optimally adaptive action-output could be implemented in a timely fashion, that is, extremely swiftly. It takes time for the brain to construct models. In the case of vision, the brain requires approximately 80 to 400 milliseconds, depending on the type of visual stimuli.

Considering that mental models act as predictors, processing speed can make the difference between life and death. Interestingly, mental models for moving objects are processed faster, which does make sense if you think of a rapidly approaching predator.

Let us consider this obscure sentence: between sensors and effectors, the in-between becomes sense, trying to make sense, compelled to make sense. Let us focus specifically on "the in-between becomes sense". This segment refers to the fact that our brain builds predictive mental models constantly without the need of sensory input or motor output. We ruminate, we worry, we daydream, we think almost constantly, well some of us more than others. This "in-between becomes sense" has taken over the original definition of sensing, that is, using sensory cells. We "sense" without our senses, and that means we can drift almost endlessly. In order to survive, species are threat detectors. Predictive mental models are at the very origin of why we can ask, "What is the meaning of life".

To calm your drifting mind, you can connect to the primary sense of sensing, which means, "direct sensory stimulation", with awareness. Observing your breath is one way. Connecting to the physical sensation of where rhythmic breathing feels more prominent: the air passing in and out through your nostrils, or your chest or belly inflating and deflating. Hence, this is why the awkward sentence ends by: let go, breathe, and "come back to your senses". Another manner of connecting to your primary senses is by stretching, or yawning. We tend to do both simultaneously. According to Bruce Hood, such action connects us back to the feeling of self; we appropriate the image of who we think we are (Bruce Hood, The Self Illusion).

We have evolved to detect threats, real or false. When mental models can be generated without actual sensory inputs, they are free to model threats (real or false) or to model what seems mysterious, or what does not seem to make sense. Sense, senses, sensing; it is all about sense, senses, sensing and their many definitions, is it not? Here are some mysteries: why are we here on earth, where do our thoughts come from, what is the soul, what is the mind, etc.?

We can use our senses to feel our body moving when we willingly or sometime unwillingly move it. The awareness we experience is a product of the brain, but we can establish a correlation between willing something and feeling it happening. This is all very well, but one can complicate the matter by reminding us that the feeling of willingness is also a product of the brain and that the brain needs time to generate our experience of free will. And yes, such time to generate the feeling of free will can exceed the time required to engender the action itself. There is experimental evidence that in some cases, the feeling of conscious will can occur after the willed action itself (Wenger, 2003; Harris 2012).

The best-known experiment was done by Benjamin Libet who asked subjects to lift each fingers separately at their own sequence and will, while they noted the precise time of their feeling of free will. Libet used electrodes to record "readiness potentials" in a region responsible for generating motor actions (the presupplementary motor area, which when electrically stimulated does generate motor outputs), and found that all occurred half a second before the subject's reported time of willing to move a finger (Libet et al.,

1983). In fact, already in the 1600s Baruch Spinoza postulated that free will is an illusion. In his book "Ethics" he wrote:

"the infant believes that it is by free will that it seeks the breast; the angry boy believes that by free will he wishes vengeance; the timid man thinks it is with free will he seeks flight; the drunkard believes that by a free command of his mind he speaks the things which when sober he wishes he had left unsaid. … All believe that they speak by a free command of the mind, whilst, in truth, they have no power to restrain the impulse which they have to speak."

We can feel our heart beating and we can feel our breath. We can feel pain and in most cases we can know the cause or perhaps ask a health professional to identify the cause. Our brain produces all of these sensations and it makes sense to us because these "feelings" are concrete events, which we can observe ourselves with our senses: we can use our five senses to feel and associate these sensations with a stimulus, an origin, and a cause. We can appreciate the smell of freshly cut grass when looking at someone mowing it. It does indeed make "sense". However, we cannot feel the activity of our neurons firing. Our brain does not generate a conscious sensation of neuronal activity itself, because (here is the infinite loop), these very neurons are the ones that generate a conscious feeling. Everything we feel, we experience, and we do consciously or automatically, requires neurons. Again, an important take home message is that neurons cannot generate a conscious experience of themselves firing.

When you talk or sing, you do not need to deliberately activate each of the many processes that take place in generating sounds; this is a relief, otherwise it would be a tedious task. The same goes with thinking, ruminating, you do not feel the neurons firing, you do not have to activate these neurons one by one, let alone you have no direct control over them, you do not even feel them. The neurons that interpose themselves between your senses and your actions or intents or thoughts, do not elicit a conscious sensation. Other types of cells are also involved: glial cells. And it is not only about neurons firing but also about oscillations in cell potential and many other events. You are oblivious to all these events, which researchers are still elucidating.

Thoughts therefore appear to come from a void, from a mysterious place. You cannot stand not knowing, you need to know where thoughts, the mind, the soul, come from. You attribute an external origin, outside the body or deeply rooted in quantum physics or string theories or any other new physics that can provide a haven for the origin of the mind, the soul. And you ask why, why are we here, what is our purpose, why do we die, what is the meaning of life?

Evolution has led to species that create predictive models of their internal and external environments, under the pressure to survive and reproduce, allowing genetic information to be exchanged, which is the only way evolution took place and still does. Evolution has no purpose other than being a passive outcome of genetic information being passed along and between species (vertically and horizontally). We are gene carriers. This is it. Or is it really, it?

The need to find meaning is related to our tendency to extract signals from noise, to obtain sensory meaning from randomness or from physical modalities that are unrelated to our need to survive and reproduce. What do we do with our capacity to form so many mental models that are not required for our survival and reproduction? We create, we make art, we write poetry, we interact with each other, we invent fictive worlds, and we exert compassion. At least we have the potential as a society to do so. We also have the potential to annihilate many species including humans; technically we have reached that point.

Our human ancestors learned that as a group that they could hunt bigger and faster animals than themselves. In the African savannahs, our ancestors formed tribes and learned to cooperate. As groups became larger, members started to rely on myths, beliefs, to create a social glue, to have something in common that would unite them, make them unique. Beliefs provided norms and values that were collectively shared within the larger tribes. One of the founders of sociology, Emile Durkheim, referred to communal norms and values as promoting a sense of belonging and of collective consciousness. The number of large tribes increased, each forming and developing their respective beliefs, something artificial that they could emulate among themselves, which eventually led to different languages, cultures and religions.

The need to give answers to existential questions, and the manner in which these questions were answered depended first and foremost on tribalism, and today we still witness the best and the worst outcomes. According to Chris D Firth (based on his studies of interacting minds), culture creates us and we create it.

Therefore, in looking for answers to questions such as "what is the meaning of life", the answer is deeply rooted in the cultural context at which time the question was addressed.

Let us travel to 19th- and 20th-century Europe, and meet with a particular "tribe" of philosophers. We bump into one of them in a café in the Paris' neighbourhood of Saint-Germain-des-Prés. She or he tells us, while dirking absinthe, smoking Gauloises and exhaling in our face, that literally life has no meaning. Then what? We get that stance from reading existentialists such as Alberta Camus (The Myth of Sisyphus; The Stranger), Jean-Paul Sartre (Being and Nothingness; Nausea), Simone de Beauvoir (Les Temps Modernes), Richard Wright (The Outsider), Friedrich Nietzsche (The Antichrist), Søren Kierkegaard (Fear and Trembling; Either/Or), Karl Jaspers (Reason and Existenz), Lev Shestov (All Things are Possible: Apotheosis of Groundlessness), Franz Kafka (The Trial), Eugène Ionesco (The Bald Soprano), and Martin Heidegger (The Nothing). As with most of these intellectuals, we might conclude that life is absurd. Are we just like Sisyphus, condemned to rolling a huge boulder endlessly up a hill? Sorry for the depressing thoughts.

Does it have to be so? If life has no meaning, where do we go from here? Compassion would certainly have its place alongside such statement. This chapter is not over, keep reading please. There is so much more than just saying life has no meaning. Otherwise, why would I write this book? Indeed, why am I doing this? Why? There is frustration when writing a book. Who will read it? This book is one of so many that addresses the meaning of life. It feels pompous to even go there; why should I write about it while so many others have pondered and offered answers? Even when presented with facts, we do not grasp them naturally; it is not an assumption one can make, it is not a given. We are biased, we protect our own vision of the world, our own creation by our brain, genetically programed to develop in a specific manner and to respond to environmental cues also in specific manners depending on our genetics-epigenetics and on our societal and cultural upbringing. So why did I write this book? It is an exercise, a process. Approaching over 50 different publishers until one might show an interest is also quite the process; it does test one's resilience.

Reflecting on the fact that life has no meaning ought to be depressing. What is the point of living, really? Why should I keep trying to get this book published after numerous refusals? Sorry about the tone here. The tone sounds very much like a threat. It is threatening to think in this manner. One experiences a feeling of void that can be debilitating, as fear detectors. But it is a false fear, it is not life threatening. Well, ultimately, life is fatal, at the end. But, meanwhile, one can live fully.

Besides the feeling of threat, something else is going on: the lack of reward. Throughout evolution, we have developed reward systems to support survival and reproduction. Reward is associated with evolutionary ancient aspects of the brain, known as the "limbic system". This part of the brain, called the ventral tegmental area (VTA) is located under the cortex and deals with emotions. A key molecule mediating the various emotional states we experience is dopamine (you could have a glance at figure 28 for a simplified version of the neural networks involved in the reward system). The evolutionary more recent part of the brain, the medial prefrontal cortex (abbreviated as mPFC in figure 28), exerts a certain level of control on our emotions (see the negative sign, signifying inhibition, on the dopamine neurons inside the VTA, which receive inputs from the mPFC). The medial prefrontal cortex and its connections to other brain regions are one of the latest aspects of the brain to fully develop as we grow up. We can picture babies as being essentially driven by emotions, and teenagers having their share of experimenting with raw emotions and at times maladroit attempts at communications between their medial prefrontal cortex and their limbic system. Pathways between the medial prefrontal cortex and the subcortical regions mediating emotions can take up to 25 years of age to be fully developed. One of the latest maturing features is the coating of axons with myelin (white matter), which expedites communication.

Back to threats. There are no rewards associated with feeling threats. But there can be a feeling of reward when being able to successfully deal with threats, by predicting key aspects of our environment ultimately through our senses and acting accordingly. So what we feel — modulated by physical factors or through intrinsic emulations independently from any physical elements — and how we respond, will determine whether dopamine is secreted as a form of reward. We can bypass the reward pathways by taking "dope" such as opium, heroine, synthetic opioids, etc. Regardless of the origin, during these short moments, nothing else matters, there are no threats; there is only elation, bliss. Reaching such states in these very manners is highly addictive, to a point that requires more frequent and higher concentrations of external drugs or compulsively reengaging in addictive behaviours such as drinking, gambling, online pornography, social media, etc. The societal damage can be devastating.

There are other means of experiencing powerful reward responses. Evolution has banked on reproduction and the associated reward, orgasm, which has proven to be most adaptative. Orgasm is the ultimate letting go. "*Sexual relief is the ultimate form of mysticism available to most people*" (Hanif Kureishi, Intimacy). At the coital climax, you accomplish (in theory) the most pertinent aspect of evolution: passing genetic information. We are talking about the species that rely on sexual reproduction to pass genetic information. This form of gene transmission is relatively recent, in fact extremely recent on the evolution's timeline.

When a state of bliss or enlightenment is reached, religious or not, distinct regions of the brain are activated as supported by functional brain imaging techniques (Newberg and Waldman 2016; Beauregard and O'Leary, 2007). There is a temptation to attribute these regions of the brain that "light up" when praying God, as proof of its existence. The fact is that the brain itself generates the concept of God, and to this mental model, one adds the state of bliss achieved when immersed in authentic prayer and contemplation. Together, these factors underlie the activation of clearly circumscribed brain regions.

In brief, beliefs in external agents such as God or Gods or other entities, can act as stabilizing beacons, with concrete measurable changes in brain activity, much alike mystical signatures. When someone reaches a state of inner peace, what is the point in telling her or him "God is a delusion"? The key question that is perhaps less raised is: why do we have the urge to believe, let alone why can we believe?

Religion transcends the reality we cannot fathom. How can all this have happened? It must have been created. There has to be a God, or some external agent, whatever you want to call it; how about "Nature" as Einstein and other scientists called it and still do? There has to be something more than us (non-materialist neuroscience).

The belief in God is a response to the lack of conscious awareness of the neuronal circuits that generate mental models. In fact, the belief itself is a mental model generated by these very neurons. The apparent endless loop — or vortex as referred to by Rodolfo Llinas in his book "The I of the Vortex" — where the observer and the subject are the same, that the action of observing is being observed itself (mirrors facing each other, creating infinity), is a false issue, a seductive trap. Focusing, even obsessing on what seems an infinite loop can constitute a powerful threat. But it is a trap that one can escape by putting in context the actual evolution of these neurons, over 3.8 billions years... a frame, a scale, that is beyond our perspective when being busy with surviving and reproducing... Now might be an opportune moment to listen to "Spiegel im Spiegel" by Arvo Pärt. Breathe again, breathe into the feeling of infinity, let go, conceptualize 3.8 billions years and a mirror looking at a mirror. Enjoy.

We are looking for external agents. The ability to do so is an exaptation of the generation of mental models by our brain. Pseudoscience is a candy store for external agents. Several studies demonstrate that people who are willing to accept that their life is affected by external (non-reality based) forces tend to believe in pseudoscience (Newby and Davis 2004; Hergovich et al. 2008; Sagone and De Caroli 2014).

Even science itself is not immune to providing external agents. Take quantum mechanics approaches to consciousness. The external agent is located in a scientifically documented process. Drs Eugene Wigner, David Bohm, Roger Penrose, Stuart Hameroff, Karl Pribram, Henry Stapp, David Pearce, Friedrich Beck, Michael Egnor, Paul L Nunez, and John Eccles are examples of well-established scientists and philosophers who attribute consciousness to quantum brain physics and new physics (Beck, 2001). Positing that consciousness entails quantum mechanics has commonalities with pseudoscientific movements that rely on various phenomena such as "nonlocality" and "the observer effect" (Heisenberg Uncertainty Principle) to support beliefs in the supernatural including life after death.

Some scientists not only put forward that consciousness functions beyond the confines of the physical body, but that it survives beyond death. With such reasoning, near death experiences support their argument. For instance, the Division of Perceptual Studies (DOPS) is a university-based research group that studies phenomena related to consciousness "*clearly functioning beyond the confines of the physical body*", as well as phenomena that are directly suggestive of post-mortem survival of consciousness. It is not because countless books have been written and numerous TV programs and documentaries provide "evidence" for life after death that there is scientific proof. Death is terrifying for many and knowing that there is something after, ought to be comforting. How about a dose of dopamine? Examples of hope abound: "Life After Death, Powerful Evidence You Will Never Die" in which the author Stephen Hawley Martin lays out "*... indisputable evidence the brain is not the source of your consciousness. Rather, it is a receiver that connects and integrates it with your body, and when the body dies, consciousness is released.*" You could doubt his credentials, knowing that he also wrote a book entitled: "The Secrets Of Successful Entrepreneurship: Start a Business, Grow a Business, Sell a Business". Perhaps you would have more trust in the internationally acclaimed neurosurgeon Dr. Eben Alexander who shared his own near death experience in his book "Proof of Heaven: A Neurosurgeon's Journey into the Afterlife". Dr. Alexander recounts waking up after near death, assured of a life beyond death, and equally confident that his brain was incapable of fabricating the journey he experienced. Feel free to search for more examples of well-established authors. They do indeed abound. The fact is that we will all die at some point, and this is it. However, we can live a humble, mindful and yes, sacred life. As I mentioned before, reductionism does not mean to desacralize; quite the opposite, it means to be respectfully curious and to have the utmost respect for the world outside and inside us. One needs to be brave to accept the reality that there is nothing after death.

"*Because our living matter is greater than we are.*" Clarice Lispector (The Hour of the Star)

Death: the ultimate threat. Why do we die? We die because we live. Why do we live? We live because we are the consequence of DNA being passed on through survival and reproduction of species (at least when relying of sexual reproduction to do so). Therefore once passing of DNA has been achieved, there is no evolutionary pressure to live longer. In some species, longer survival of the parents is favourable when the progeny needs more time after birth to become fully independent for its own survival. Once there is no longer any evolutionary pressure to survive, aging processes can begin. In human, this means, approximately around age 25. You will of course not notice any obvious aging traits typified by wrinkles until the mid forties (neck and hands are the most obvious sites); we mentioned in the first chapter, the progressive difficulty to focus near objects, which occurs in the mid-forties. Therefore, how does the aging process manifest itself so early in life, in the mid-twenties? You might have heard of the shortening of telomeres. Forming the ends of human chromosomes, telomeres shorten every instance a cell divides, which therefore limits the proliferation of human cells to a finite number of divisions. This is a very slow process that affects – obviously – only the cells that divide, such as skin cells. More recently, another process has been proposed. It has to do with protein complexes involved in maintaining the quality control of all proteins (one wonders how they maintain their own quality). Since proteins are essential for cell survival, protein quality control processes (known as proteasomes) affect all cells of the body. Proteasomes will eliminate unneeded or damaged proteins by breaking them down (proteolysis). Other players that are absolutely essential in ensuring protein quality are chaperones and co-chaperones. Aging starts insidiously with the function of the proteasome and chaperones and co-chaperones becoming compromised (Kelmer et al., 2020; Margulis et al., 2020). We can be jealous of species that have exceptional regenerative capacities, and become in fact healthier with age: hydra, jellyfish, clams, lobsters, turtles, flatworms, Greenland sharks, bowhead whales, water bears (tardigardes) and certain bacteria. In 2015, nematodes were restored to life after having been buried for over 32,000 years near the Alazeya River in Russia. This beats by far the record of the officially documented oldest human, Jeanne Louise Calment (21 February 1875 – 4 August 1997) who lived for 122 years and 164 days.

In conclusion, we have evolved to die after genetic information has been passed and can be further passed on by our progeny. We start aging approximately when our progeny has become autonomous. Once we understand the context, we can allow our life to be fulfilled and sacred. We can approach death with poetry; we can find beauty and we can create it. Here is Mary Elizabeth Fry's gift to us in only twelve lines expressed as iambic tetrameters, that is, four feet (8 syllables) per line, with the exception of lines five and seven: Do Not Stand at My Grave and Weep.

Do not stand at my grave and weep
I am not there. I do not sleep.
I am a thousand winds that blow.
I am the diamond glints on snow.
I am the sunlight on ripened grain.
I am the gentle autumn rain.
When you awaken in the morning's hush
I am the swift uplifting rush
Of quiet birds in circled flight.
I am the soft stars that shine at night.
Do not stand at my grave and cry;
I am not there. I did not die.

Life and death remain mysterious for many and it is a completely understandable longing to try finding meaning in external agents. Highly respected cognitive scientist Jerry Fodor was a strong proponent of external agents. He maintained, "*Higher-level theory of psychology or linguistics could not be captured by the low-level explanations of the behaviour of neurons and synapses*". Interestingly, Fodor opposed Darwins' theory of evolution. In fact, it is by following evolution's trial and error that we can appreciate the continuous processes of sensing and acting for survival and passing of genetic information. These very processes led incidentally, or one could say accidentally, to the interposition of neurons between sensors and effectors, generating a predictive mental model of the external and internal world in which species continue to evolve. One must avoid being anthropocentric; we are not a purpose but rather a consequence of 3.8 billion years of evolution.

Some scientists manage to reconcile evolution with the belief in God. Take Francis Collins, who headed the human genome project. When discussing with evolutionary biologist Richard Dawkins about the proof of God's existence, Collins argued, "*One couldn't rule out anything that might be outside of the natural world, such as God*". You cannot rule out God with science (we will get back to this argument just below). Collins continued, "*... your mind must be opened to aspects you cannot grasp with science, or that are outside of the physical world*". On a long drive with The Skeptics Society founder Michael Shermer, Collins shared that whenever he experienced doubts, and he still does, he saw it as an opportunity to "continue growing" (The Believing Brain, Shermer M). What I find fascinating is that according to Shermer, Collins does not need God to explain his scientific pursuits such as understanding evolution and concluding that mouse and man share a common ancestor. Shermer proposes that Collins' belief has deep emotional roots. Having seen his patients suffer as a resident and the role their faith played, Collins was emotionally shaken, which led him to read and become inspired by the writings of CS Lewis, who himself was a non-believer who eventually "found God". Walking in nature, exhausted from overworking as a medical doctor, Collins also "experienced the presence of God" in that one instance.

Therefore it is possible to reconcile accepting evolution with being a dualist such as believing in

God or the supernatural. What about the continuous process of embryonic development? Another reflexion point, which is also factual, is how every human being is born out of a sperm fertilizing an ovum (egg cell). As is the case with evolution, embryonic development does not require anything outside of the domain of the physical. The fertilized ovum becomes a zygote, which then cleaves to form a morula, then a blastula that will anchor itself in the endometrium of the uterus to continue dividing, then forming a gastrula featuring three germ layers, each leading to specific tissues, finally the nervous system begins forming followed by all other organs. Development continues after birth, with the latest organ to be fully mature being what? You guessed it... the brain. As we mentioned above, full maturation is estimated to be completed as late as 25 years of age. Again, the process is continuous, no external agents involved, and humans are only one example of the fact that reproduction in all multicellular species is followed by well-documented developmental stages. Of interest are the similarities of these stages and related genes and mechanisms between species, such as mouse and human. In fact, the mouse closely mimics human development from zygote to sexual maturity, hence its pertinence in scientific research. Zebrafish are also highly pertinent models of early human development and physiology.

Science is the best tool we have to grasp the world we live in. Using the scientific method of experimentally testing hypotheses, the boundaries of beliefs shrink: the earth is not the centre of the universe, the earth is older than six thousand years, evolution is a fact, we are oblivious to our thinking processes and these are very much part of our body. The beauty is that scientific observations remain disprovable if new experimental evidence refutes them. Science is a process. Beliefs tend to reflect the time of history when they arose, accounting for the scientific observations available at that period. Flat Earth societies created in the 20th century are one of the exceptions that confirm the rule. Some members are not serious, but others, such as conspiracy theorist Nathan Thompson, believe that the Earth is a disc with the Arctic Circle at its centre, while Antarctica is a 150-foot-tall wall of ice, around the rim, guarded by NASA employees who's sole purpose is to stop people from climbing over and vanishing into space. Here was a nice digression. This belief is easy to refute. How about abstract concepts?

A key issue is that we humans create concepts such as the soul, consciousness, God, which become entities of their own. As entities, the soul, consciousness and God become the subjects of studies. We can now devote our life to studying the concepts we created from scratch. Philosopher David Chalmers canned the term "hard problem of consciousness", which he claimed would be unsolvable even if we could at some point exhaustively explain the specific functions accomplished by the brain. Believers will continue to believe unless they are provided with scientific evidence that refutes their beliefs. Do you see the problem here?

One creates an abstract concept that has no bearing with reality and expects to find evidence or counter-evidence in reality. A false debate is created. If we cannot find evidence within reality, then these concepts must belong outside reality; it has to be, end of discussion. Who are we to think we know it all. My stance is that one way out of this trap is to understand the origin of such concepts: 1) why we created them at the first place; and 2) how were we able to create them? There is one possible answer and we talked about it again and again. Observe, using science, the evolution of predictive mental models that are independent of sensory input and motor output. These models have "a mind of their own". They are the very foundation of why we ruminate. One argument to fuel beliefs is that science is not advanced enough, or that science cannot capture all that exists. This is very much the stance of mind-body dualists. You cannot disprove that the mind exists independently from the body, from the physical. But you can ask how we came about to create this concept, in the context of sensory evolution. In theory, we can venture there with humanity and compassion, without representing any threats; in theory. For many devoted scholars, it is hard to get the mind out of their mind, to get consciousness (as an entity) out of their consciousness (as a predictive mental model).

Take philosopher Philip Goff who devoted his career to studying consciousness as an entity. He states: "*Nothing is more certain than consciousness, and yet nothing is harder to incorporate into our scientific picture of the world.*" (Consciousness and Fundamental Reality, 2017). Goff proposes that we need to revise our view of the physical world to accommodate consciousness. He put forward panpsychism, which means that all matter is conscious, that consciousness is everywhere. In his view, matter does not suddenly become conscious by re-arranging itself; instead, consciousness is there from the start. Goff refutes mind and brain dualism by proposing that subatomic particles such as protons and electrons all possess a level of consciousness and that by experiencing change, more complex forms of consciousness can be built up from the simpler ones. It ensues that all that we human can sense (the colour red, the smell of coffee, a loving touch) have their very foundations in matter, which already possesses consciousness. While trying to unify mind and matter, an external agent is introduced. Consciousness builds on intrinsic properties of subatomic particles that possess a basic level of consciousness. According to Goff, "…. *the intrinsic nature of matter inside of brains, in the sense that both inside and outside of brains matter, has an intrinsic nature made up of forms of consciousness.*"

At the risk of repeating myself, the evolution of *Homo sapiens* is incidental (not a goal) of trial and error in passing genetic information that allowed survival and reproduction, and this took 3.8 billion years, with 86% of this time spent as single cells evolving. The physical world we can fathom is limited to the physical modalities that had to be detected to ensure the survival and reproduction of our ancestors and distant cousins, and much more recently to the technology we have developed and still do. We lack the imagination to be humbled by how limited our representation is of the physical world we live in. Both space and time are an infinite continuum, with no beginning, and no end. We do not "see" atoms, we do not feel the distortion of time by gravitational forces… we do not "see" any of the 4 forces we have identified… we know them because of the tools we have developed.

It is tempting to attribute external agents to the new dimensions of reality that were opened to us by the technology we developed and still do. According to Goff, consciousness as an entity is already present in subatomic particles. For some, quantum mechanics can explain the mind. How about superstrings? They are vibrations, we are vibrations, and the soul is vibrations, and the souls of all humans and the entire universe are connected through these vibrations. There will be more discoveries in physics, alongside new explanations for the entities we created using our mental models, to justify that consciousness, mind, soul, exist independently of the body as study subjects.

Friction even hostility is bound to arise when arguing whether body and soul are distinct, whether there is something more than the physical body. Guess who was sitting beside pseudoscientist Deepak Chopra (Channel 4 UK documentary The Enemies of Reason, 2007)? Richard Dawkins, again on a mission to prove that God does not exist. The discussion turned into almost hostility. The scenario was similar when Chopra and Jean Houston (co-founder of The Foundation for Mind Research) debated neuroscientist Sam Harris and sceptic Michael Shermer on the question "Does God Have a Future" at the California Institute of Technology in 2010. Imagine, had Bertrand Russell participated in such a debate. He might slightly have inflamed the tone further by saying something in the likes of "*Religion is based ... mainly upon fear ... fear of the mysterious, fear of defeat, fear of death. Fear is the parent of cruelty, and therefore it is no wonder if cruelty and religion have gone hand in hand. My own view on religion is that of Lucretius. I regard it as a disease born of fear and as a source of untold misery to the human race.*" For the cherry on the sundae, we would also have invited Blaise Pascal, who might have concluded: "*Men never do evil so completely and cheerfully as when they do it from religious convictions*".

I had my share of hostility in the most unexpected moment. Riding my bicycle up and down a hill, I came across a man walking. After 20 encounters, I stopped and engaged in a conversation, by invoking the myth of Sisyphus. He replied that there was nothing absurd about going up and down the hill. It was a hot summer. The ice was broken. I told him about the book I was writing, briefly stating the evolutionary continuity of sensory input and output in terms of the cells interposed between them, to the point that these cells could create a predictive mental model without input and output. I also told him that we could not sense these cells located between sensors and effectors, which can constitute a mystery, a void, and require a need to rely on external agents. He told me he disagreed, which made me want to listen to his point of view. He said that all begun with consciousness, that I had it all upside down. He also said I was not truly listening to him, that I was arrogant and even a threat to humanity by ignoring the documented spiritual revelations of so many great thinkers. What had just happened was exactly what I wanted to avoid. No doubt he had issues with me, on my bicycle beside the road, trying to have a conversation; and yes I might have shaken my head a few times while he was talking. He appeared very angry, started to walk away, saying he never wanted to see me ever again, also assertively stating I had no humanity. All I could reply as his silhouette shrunk in my field of view was, that I was not angry. The practical issue is that I very much enjoy cycling up and down that hill, conveniently located behind my house. Shall I change my schedule to avoid another encounter? Perhaps alter my physical appearance or wear a full-face helmet and cover myself with a drape, which would look dubious on someone riding a road bike, unless it were Halloween, which would limit the occasions to ride... A humbling experience for someone who is adamant that reductionism is compatible with humanism and compassion. It does take two to tango, so I accept my share of blame.

In his book "The Meaning of Human Existence" ant biologist Edward O. Wilson says, "*The great religions are also, and tragically, sources of ceaseless and unnecessary suffering. They are impediments to the grasp of reality needed to solve most social problems in the real world. Their exquisitely human flaw is tribalism.*" He adds that leaders of each religion and sect should "*publicly defend the supernatural details of their faiths*" and if they fail to do so, he recommends that we "*charge with blasphemy any religious or political leader who claims to speak with or on behalf of God.*" His answer to the meaning of life relates to our evolutionary development and what we make of our future. He states: "*We are all genetic chimeras, at once saints and sinners, champions of the truth and hypocrites — not because humanity has failed to reach some foreordained religious or ideological ideal, but because of the way our species originated across millions of years of biological evolution.*"

I agree that one must absolutely consider the context of evolution; however, the challenge as a society is to educate with compassion. Call me a dreamer, an idealist, even call me naïve, but I do think the greatest challenge is to compassionately educate humans so they learn facts about how we evolved and how our brain works, among many other fascinating aspects of our reality. There has to be a way to teach everybody without threats. Who reads books like this one? Just very few; sorry, dear publishing house. We need more of those passionate science teachers. Well, we need more teachers in general, so that facts can be presented without hostility by role models, and when possible in a playful fashion. Let us honour the annual winners of the Varkey Foundation Global Teacher Prize of one million dollar: Ranjitsinh Disale (India) 2020; Peter Tabichi (Kenya), 2019; Andria Zafirakou (UK), 2018; Maggie MacDonnell (Canada), 2017; Hanan Al Hroub (Palestine), 2016; and Nancy Atwell (USA), 2015. Once parents learn, they also can help their children to learn and to be more receptive in class and in life in general. We need to foster curiosity. On why and how we should ultimately educate, Steven Pinker states: "*The goal of education is to make up for the shortcomings in our instinctive ways of thinking about the physical and social world. And education is likely to succeed not by trying to implant abstract statements in empty minds but by the taking mental models that are our standard equipment, applying them to new subjects in selective analogies, and assembling them into new and more sophisticated combinations*" (Pinker, 2004). Education in its largest sense is part of the solution to getting along.

"*[Education] seems to be the only hope: to spread enlightenment sufficiently wide so that people can no longer be manipulated to give consent to policies that will surely eventually lead to their own destruction.*" Anatol Rapoport (The study of conflict, 1968)

What happened to humanism? Can we get along as adults? In trying to do so, I plead guilty. I did qualify Chopra as a pseudoscientist, which likely would have led to similar non-constructive cul-de-sac outcome; I also looked up his estimated net worth, 150 millions USD, and compared it to Dawkins' at 10 million, which is still up there. Back to Russell, he does inspire me; I also draw pleasure from reading Kant and Montaigne among others. Oh yes, you might have noticed that I am nourished by poetry, one of my favourite poets being Clarice Lispector.

One can use poetry while avoiding the so-called "pathetic fallacy" (a name given by British cultural critic John Ruskin) where human emotions, intents, or conscious awareness are attributed to describe the natural world; if ever doing so, one must be transparent of her-his intent. For instance in the first chapter's ode to transduction, the poem is obviously intentional in its prosaic form, while the scientific translation carefully avoids "pathetic fallacy"; the outcome is awful from a poetic point of view.

Back on track, we need more poetry in our world, no seriously really back on track. Here is what I propose: compassion as opposed to confrontations and debates on whether God or any external agents exist or not. Confrontations and debates fail the point as a society to evolve as a group. One can use compassion even with oneself (Hofmann et al., 2011). For instance, you read this book and you believe in God, in dualism, in external agents or you doubt (agnostic). Then, you have kindness for you, you understand that ultimately what matters is the outcome on your take on life, your profound desire to seek, to embrace what appears to be greater than you, what you cannot comprehend. You feel bliss, regardless of how; you imagine the specific regions of your brains that are activated. In my own moments of bliss, I very much share the same regions... the same territories where we can meet. Don't we all want to feel bliss? This book and the author are not a threat to believers. We all ultimately thrive from inviting the sacred in our lives. Why did you pick this book? Why did I write it? For the same reasons, I very much hope.

The book you are reading now is an attempt to explain the roots of such spiritual voyage, to reach for the infinite, to be greater than oneself, to experience bliss, to live our spirituality. From its very title, this book aims at giving a meaning to life, a reason to thrive, to live fully, to embrace life. Simply saying there is no God could be a threat or be irrelevant to a believer at peace with herself-himself-themselves. Realizing there is no God is not where it stops, it is where it starts in your journey, love oneself, be in acceptance, observe what we know.

I, the writer of this book, embrace the quest for a spiritual life without requiring external agents. This in no way precludes me from having compassion for believers in God, believers in a spirit, mystics, and agnostics (those who do not pretend to know). My label is "atheist" (there is no God), and this does not define who I am as a human being, certainly not. Therefore, I rarely use this label unless the context asks for it. I am offering links between established observations, facts. I am not a threat to believers; here is an opportunity to "look" at your beliefs in a more encompassing context. When addressing the nature of consciousness, do so by respecting what we can observe scientifically, what still holds at this point in time, i.e. what has not been disproved by observations. Again, consciousness is not a thing and therefore cannot be studied as an entity. Consciousness is a state of experiencing predictive mental models, linked or not to your senses and outputs.

From a medical point of view, especially if you are an anaesthesiologist, or are the patient about to undergo invasive surgery, you will want a definition of consciousness; you will want to make sure that anesthesia will induce a state of "absence of consciousness". We can all agree that definitions of states or levels of consciousness have nothing mysterious. There is no need for any external agents here. Chemical substances can induce the loss of consciousness, and they unlikely act on any external agents.

Briefly, let us consider how we can study consciousness in humans while acknowledging that it is not an entity or not anything mysterious or metaphysical, and certainly that it does not involve external agents. I mentioned that consciousness was a state: we are awake or in a coma, or at graded levels in between those two states. This is one aspect we can clearly study: the level of consciousness. According to experimental psychologist and neuroimaging expert Chris D Frith (who published over 400 peer-reviewed scientific publications), we can study two additional aspects. The second aspect represents the contents of consciousness: what determines how a small amount of sensory information is associated with subjective experience, while the rest is not. The third aspect is meta-consciousness, the ability to reflect upon our subjective experiences and, importantly, to share them with others (Frith, 2019). Considerable advances have been made on the intricately detailed mechanisms that respectively underlie these three aspects of consciousness.

Let us start with the neural basis that controls the level of consciousness. Quoting Michael S Gazzaniga (2018), at our most awake level, "*It's a summation of processes that all happen in parallel and our consciousness is a product of these interacting parts.*" Consciousness is a process that involves complex dynamic interactions of multiple brain regions, including the brain stem, midbrain and cortex (Demertzi et al., 2019). Specialized regions of the brain are activated together and through time they are "stitched" together to give the illusion of a unified consciousness. Not surprisingly, consciousness involves long-range functional connections and synchronous oscillations between remote regions of the brain.

Key distant connectivity and synchrony occur between thalamo-cortical loops and large-scale cortical networks that link the frontal and parietal cortices (plural of the word cortex). The model putting forward that the recruitment of frontoparietal networks are essential to generate a subjective experience, is known as "global neuronal workspace" (GNW) (Dehaene et al., 1998). According to the GNW model, different streams of information in the brain compete for the global activation of a widespread network of regions in order for a subject to access consciousness. More recently, Dinesh Pal and colleagues (2018) obtained experimental evidence from anesthetised rats using selective reversals in specific regions, that the prefrontal cortex does play a role in controlling the level of consciousness. Finally, studies using attentional blinks and transcranial direct current stimulation — to potentiate or reduce neural excitability — showed a role of the dorsolateral prefrontal cortex in modulating visual awareness.

Under a coma state, these widespread connections and activations are lost, which impairs the integration of information between distant brain regions. Brain metabolism is also reduced, particularly in the frontal and parietal cortex, by at least 40%, which contributes to the loss of functional connectivity between remote brain regions: the integration of large-scale information is compromised. Deep general anesthesia also induces similar losses in consciousness, without interfering with metabolism. There is experimental support that communication between the thalamus and the prefrontal and anterior cingulate cortices contributes to consciousness. Electrical stimulation of the thalamus was associated with behavioural improvements in minimally conscious patients afflicted with severe traumatic brain injury (Schiff et al., 2007).

Another way to "link" distant brain regions is to have neurons synchronizing their spontaneous oscillations, i.e. depolarizing at overlapping frequencies. High frequency synchronizations (> 40Hz) link neurons locally while low-frequency activity is involved in information processing between distant brain regions; this phenomenon is known as communication through coherence. With the electroencephalogram, one can measure three bands of slow oscillations: theta (4–7 Hz), alpha (8–13 Hz), and beta (14–25 Hz) (Bonnefond et al., 2017). Other key players in regulating the level of consciousness are three distinct types of GABAergic interneurons that do so by modulating the dynamic reorganization of functional networks (Modolo et al., 2020). Of note, levels of consciousness entail a great deal of neuronal circuit plasticity. There is one more region we need to talk about, the reticular formation. Located at the base of the brain, within the brainstem, this region sends neural projections to the thalamus and cerebral cortex that modulate which sensory signals reach the cortex to elicit a conscious percept. In addition, the reticular formation plays an essential role in controlling the levels of consciousness. Lesions to the reticular formation have been associated with irreversible coma.

The second quantifiable aspect of consciousness is its contents: what type of mental model is the brain generating? This concept is known as the "global workspace theory". Specific mental contents — perceptions, thoughts, or intentions to act — reach consciousness when attaining a "global workspace" that broadcasts specific contents throughout the brain. In this manner, an individual can adapt her-his behaviour according to ever changing environmental contexts.

While the ultimate feat of the global workspace is to provide the illusion that all sensory modalities are one and only, researchers have relied on visual tasks to yield insightful data. Let us consider a well-established experimental design: binocular rivalry. Subjects are presented with a face in one eye and a house in the other eye. They do not see a mixture of a face and a house (try it). Instead, conscious perception of the face and house alternate every few seconds while the sensory input remains the same: a face in one eye and a house in the other eye. How do we know that? Volunteers are asked to press one button when seeing a face and the other when seeing a house. In this manner, brain activity shifts to match the respective contents of consciousness. Processing a face and a house involves two distinct pathways. Consciously seeing a face (generating its percept) is associated with increased activity in the face area of the fusiform cortex. However, generating the percept of a home engages the place area of the para-hippocampal cortex.

Now we can have some fun; we can use transcranial magnetic stimulation to selectively activate either of these two regions. Doing so in healthy volunteers alters both changes in detection (Beck et al., 2006) and binocular rivalry dynamic (Zaretskaya et al., 2010). In the first chapter we also presented an example of how the brain generates what you consciously experience as vision. The illusion in figure 19 elicits the experience of motion, while in fact the image is still. Nothing is actually moving; however, the brain region involved in visual motion detection (V5/MT) is activated. In fact, electrical stimulation of this region elicits the feeling of motion, while brain damage circumscribed to V5/MT prevents subjects from experiencing motion, a condition called akinetopsia (Zeki, 1991). To put the nail in the coffin, we also mentioned in the first chapter that experimental inactivation of V5 with transcranial magnetic stimulation induced temporary akinetopsia (Beckers and Zeki 1995).

How about contents other than those related to vision? How about your conscious experience that you willed an action? There are still intense debates on whether the experience of conscious-will precedes or is even required to perform an action. What is clear is that functional brain imaging studies pinpoint a distinct brain region: the pre-SMA (Pre-Supplementary Motor Area), which activity has been linked with awareness of intentions to act (Altena et al., 2013). Specific electrical stimulation of the pre-SMA has been associated with the urge to move one's specific body part (Fried et al., 1991). To conclude with the content of consciousness, one has to be conscious. Activity in the specific regions mentioned here also should be synchronized with the interactions between the frontal cortex and the parietal cortex as stated in the first aspect of consciousness: its level. Awareness of specific contents can also be modulated through inputs from the reticular formation.

Take vision for instance. In the first chapter we mentioned that the only way eyes could send visual information to the brain is through neurons called retinal ganglion cells (RGCs). For conscious vision, these neurons must establish contacts (synapses) with relay neurons in a region of the thalamus called the dLGN (dorsolateral geniculate nucleus), these relay neurons in turn project to the primary visual cortex. For conscious vision, this is a non-negotiable prerequisite. Curiosity arose from studying synapses in the dLGN: those between RGCs and relay neurons account for only 20% of the total synapses. What about the other 80%? It turns out that a large number of synapses represent neural projections from the reticular formation (thalamic reticular nucleus) onto relay neurons. Of interest, relay neurons also project to the reticular formation, creating a reentry loop. The word reentry means that one region affects the other, which in turn affects the other, etc., creating a loop. The reentry loop between dLGN and thalamic reticular nucleus is known to mediate visual attention. To complicate matter, the amygdala (involved with the prefrontal cortex in determining what is a real or false threat) also sends projections to the thalamic reticular nucleus. Do you get the picture? What you experience as conscious vision is somehow related to whether your brain unconsciously considers the visual stimulus as a threat or not. I will stop here, because it is a lot to take in.

However, before we move on to "meta-consciousness" I would like to revisit the part in the first chapter where we talked about an ice-breaking statement at a cocktail party: we are blind on average 4.5 hours every day. We selectively loose conscious vision for about 0.1 seconds every time we deliberately peruse the visual world around us and another 0.1 second every time we blink. Back to the cocktail party, you joined your hands together above your head to form a circle with your arched arms and asked the victim to move her-his eyes along the circle you created. Then you took turn, and your cocktail partner immediately realized the jerky movements of your eyes (these are called saccades). You explained that each saccade lasts about 0.1 seconds, during which there is no conscious vision. If you tally these 0.1 seconds for every time you peruse the world around you each waking day (a total of around 150,000 saccades), an impressive time of 4 hours is reached, on average (Melcher and Colby, 2008). You also added the 30 minutes tally of blinking; yielding 4.5 hours spent blind every waking day. We also mentioned something about a toilet, and the usual mirror found in such place. We proposed that you would stand in front of the mirror and try catching your eyes moving.

No matter how much you tried, you could not see your own eyes moving. The reason is related to saccades again. You moved from one target to another, therefore, you elicited saccades, during which you were blind, or if you prefer, lost consciousness about a specific content, vision. Your brain does not allow consciousness of your own eye movements, otherwise you would be dizzy, and you might very well puke (apologies if you eat and read at the same time, it is your right). You have no awareness of the processes that move your eyes; this is an example that the events generating eye movements occur unconsciously. Consciousness only accompanies the feeling that we "will" eye movements.

Finally, the third and last aspect of consciousness, as defined by Chris D Frith, is meta-consciousness, or being aware that we are conscious, or thinking about thinking. It can be both a gift and a curse. This is the type of consciousness that leads to questions such as "what is the meaning of life?" This third aspect allows us to think without any sensory input or motor output. Here is one example taken from vision. We just mentioned that projections from the eye via retinal ganglion cells to the dLGN neurons — which relay information for conscious perception in the cortex — represent only 20% of the total synapses. In addition to massive inputs from the reticular formation, the remaining synapses also include inputs from: the primary visual cortex itself, the superior colliculus, the pretectal nuclei, and even local dLGN interneurons. Moreover, the dLGN also receives inputs from brainstem regions that are not involved in vision but rather in arousal and emotions such as the mesencephalic reticular formation, the dorsal raphe nucleus, the periaqueuctal grey matter, and the locus coeruleus (Guillery and Sherman, 2002). Let us stop and think about this.

If you consider projections from the primary visual cortex to the dLGN, you might think this is reentry and hence, this could represent an infinite loop. Well, it is not reentry because dLGN inputs to the visual cortex synapse on neurons from different layers (4 and 2-3) than the neurons projecting to dLGN relay neurons (cortical layer 6). If you are a dLGN relay neuron, regardless of the source of your input (from the retina or from layer 6 of the primary visual cortex, or other brain regions), your role is to convey visual information to the layers 4 and 2-3 for the primary visual cortex for conscious vision. There is no way you can tell whether the input is from the eye or from the primary visual cortex or other brain regions themselves. As such you can "see" with your eyes closed, and you can be blind with your eyes opened (due to modulatory inputs from brainstem nuclei). We do it all the time; we form mental models that can be quite vivid, and yes we are blind about 4.5 hours each waking day (when your eyes do saccades and blink). Think of someone, focus mentally; can you see her or his face? But this has nothing to do with the real world. You can imagine anything. Often images appear out of nowhere. We do not know the precise circuits involved, but the fact that the visual cortex can stimulate the thalamus, which can then stimulate the visual cortex, is part of the trick. The brain can create sensory experiences that have nothing to do with actual sensory inputs.

We can ruminate, and share our cogitations with others; we can fully express our innate inclinations to be social animals. Learning and memory are major players. The prefrontal cortex is also elemental in modulating higher executive functions, such as those that allow us to be socially compliant: perhaps this is not the time to swear, or perhaps I should keep my mouth shut, and so on.

This last aspect is where we find the illusion of self, and the multifaceted personality traits that make us unique. Thinking about thinking also relates to agency: I willed this action. Conflicts arise when you produce a movement or say something without intending to do so. Agency vanishes, as the effect you intended to produce did not in fact occur. You pay a "dissonance penalty" when prediction is wrong such as in deceptions, not wanting to arbour the feelings our brain creates, taking these feelings as face values, interpreting these feelings as threats, even though they are a creation of the brain. Reacting to these feelings as if these were real threats inexorably wastes tremendous energy. One feels tired, depressed. So we find ways of overriding the dissonance penalty of energy expenditure to reinstate alignment with perception of reality, using dopamine-like inducers (alcohol, drugs, over eating, sex, gambling, anti-depressants, etc.) to activate our reward centres. All these aspects have been studied experimentally and remain an intense field of research. Specific brain regions and their interactions have been associated with the multiple features of meta-consciousness. I will not venture further, other than to remind us that there are major implications in our society. Only with self-reflection and introspection can you learn, adapt, have compassion, contribute positively to society, to your own wellbeing and to that of others.

Therefore, paradoxically, while an analytic approach to thinking might sound cold and liken us to complex machine, we are learning about our capacity for self-reflection and introspection.

Analysing does not imply, in any way shape or form, desacralizing. We need the sacred in our lives, more than ever. We must apply the highest moral standards, more than ever. As Pandora's box is revealed, possibilities are endless, for the best and for the worst. It happened with nuclear fission, it could happen with malignant artificial brain modulation.

Once we learn the facts that are available through science, in the most compassionate, respectful and constructive manner, how do we live our life with this question: what is the meaning of life?

We feel it.

Chapter VI

Feeling the Meaning of Life

While neuroscientist Sam Harris spent his career refuting religions, he has displayed over the years open-mindedness, especially mindfulness, to the quest for spiritual experiences. I found hope in his 2004 book "Waking up: A Guide to Spirituality Without Religion". An adept at contemplative meditation, Harris maintains that we can live happy lives by paying attention to the present moment. In this manner, questions such as "what is the meaning of life?" are taken in a new liberating context.

Compliant with a scientific approach, Harris reminds us that thoughts are not real, are not an entity, that thoughts have a mind of their own. He acknowledges the measurable effects that religious beliefs have on the brain. He also recognizes our urge to posit and answer the question "what is the meaning of life?" while proposing a way to live the question that has no answer.

I cannot agree more. At various places in this book, I wrote, "breathe". My rationale was that paying attention to your breath connects you directly to your senses, allowing your thoughts to live their own life, accepting that thoughts are not entities, recognising that they captured your attention away from your breath and patiently reconnecting to the sensation of your breath, again and again, a thousand times. In this manner you develop a new relationship with your thoughts, you let them be, you acknowledge rumination without reacting, you are not intimidated by your thoughts, no matter how rebarbative they might be. In the fourth chapter, I spent time establishing that thoughts are incidental to the evolution of interneurons that build predictive mental models.

I also mentioned in the fifth chapter that once these mental models evolved to become independent of sensory input and output (feeding on memories, on false fears, on dopamine as a reward), existential crisis was born. The liberated mental models acquired the ability to imprison us. Disconnected from primary senses, they became the cause of torment, anxiety, unease, and spleen; not the organ but the pain of living, as canned by French poet Baudelaire. In "The Flowers of Evil", spleen represents "*melancholy with no apparent cause, characterised by disgust with everything*". Can you beat that?

Try German philosopher and pessimist Arthur Schopenhauer. In his book "On the Sufferings of the World" (1851), he writes: "*Unless suffering is the direct and immediate object of life, our existence must entirely fail of its aim.*" Schopenhauer must have paid much attention to his thoughts, feeding on them. According to him, "*If you accustom yourself to this view of life you will regulate your expectations accordingly, and cease to look upon all its disagreeable incidents … as anything unusual or irregular; nay, you will find everything is as it should be, in a world where each of us pays the penalty of existence in [their] own particular way*". Perhaps I am reading too much into his views, but allow me suggesting that he was spot on about accepting these incessant thoughts, letting them be. Here they are, the sources of suffering, but thoughts are not real, so I return to my breath, a thousand times, I am not intimidated. Lets take a segment of the above quote: "... *you will regulate your expectations accordingly, and cease to look upon all its disagreeable incidents*".

Returning to the feeling of your breath, you connect with your primary sense of touch: feeling the air passing through your nostrils. Or you connect to another primary sense, proprioception, feeling your chest inflating and deflating. Thoughts (or "disagreeable incidents") have their own life; you get back to your breathing, to the now. This is how one can regulate. However, I would disagree with Schopenhauer that each of us should pay the penalty of existence in [their] own particular way. Let us all stop for a moment and breathe mindfully in stillness. Inspire, be inspired, let go, take the leap of faith.

"*Where there is great love there are always miracles. Miracles rest not so much upon faces or voices or healing power coming to us from far off, but on our perceptions being made finer, so that for a moment our eyes can see and our ears can hear what is there about us always.*" Willa Cather (Death Comes for the Archbishop)

States of elation, bliss, can be reached with meditation (a fancy word for observation). Meditating is observing, paying attention in the now. Contemporary examples include Buddhist monk Thích Nhất Hanh whose practice of mindfulness of breathing and of social engagement has touched many. We simply can let go, feel with our senses. In Zen practice, one lets go when hanging from a cliff. The cliff represents the illusion of threat; the cliff is a false threat.

"*It is because I dove into the abyss that I am beginning to love the abyss I am made of.*" Clarice Lispector (The Passion According to G.H.)

One takes the leap of faith and lets go off the cliff. "Coming back to your senses" in their basic sense, senses as transducers, transforming the air passing through the nose into a stable sensory experience, which in turns stabilizes the breathing itself, through the vagal nerve (parasympathetic system). Rumination follows its own unpredictable rules; however, one connects back to the breath and lets the thoughts be, the freewheeling mental models, many of which announce false threats, false predictions or these models reminisce traumatic past experiences.

At some level, unhappiness, depression and anxiety all stem from wanting to change our current state as if it were a threat. Wanting to change these thoughts, taking them for face value (they are not real), can lead to unhappiness.

"*Worry does not take away tomorrow's troubles. It takes away today's peace*" (Randy Armstrong).

Letting go is to realize there is no threat. When are we not detecting threats? There are no felt threats during an orgasm, during dopamine-loaded rewards such as winning high-stakes competitions or lottery, using opioids, or falling in love. Most anti-depressant drugs — SSRIs (Selective Serotonin Reuptake Inhibitors) and SNRIs (Serotonin and Norepinephrine Reuptake Inhibitors) — act not on the reward system but by suppressing threats in the amygdala and prefrontal cortex among other brain regions.

Another way to address depression is to be mindful of the many false threats conveyed by the revving thoughts. In his book, "The Ravenous Brain: How the new science of consciousness explains our insatiable search for meaning", Daniel Bor reminds us that the brain selects which sensory modulation to bring to conscious attention. It does so by searching for patterns amidst the constant bombardment of information. Patterns are fragmented, compressed, and made available to consciousness. For a simple example involving conscious vision, see the illusion in figure 1 where Gabor patches are combined to make a mismatch with reality. Fragmentation and compression are vulnerable processes that can lead not only to illusions but also most gravely to mental illnesses. Imagine emotionally charged memory fragments combined to construct false threats, leading to anxiety and loss of hope, i.e. depression. Daniel Bor argues that many mental illnesses could be viewed as disorders of consciousness. Mindfulness allows letting the thoughts be, without acting on them, without being intimidated by them. It is not easy to achieve a state or inner peace. Thoughts flow freely again and again, distracting you from focusing on your breath; you return again and again, a thousand times to your breath. We owe a great deal to Jon Kabat-Zinn (who studied with Thích Nhất Hạnh) for sensitizing and educating the Western world to mindfulness in a secular manner oriented to reduce stress.

Our potential to experience happiness is as infinite as our willingness, our mindfulness, to scale in the direct sensory inputs, to connect with our senses as finely, as focussed, as mindfully as possible, to connect directly to our senses. Optimal focus (awareness) or direct modulation of our brain by our senses implies to let go of judgement, to let thoughts flow freely, so that the modulation of brain activity by our senses does dominate our awareness. Not reacting to past (rumination), or future (plans, worries), just living in the present moment. With meditation (letting go, dying to oneself) or prayer (letting go to God, Nature, Spirit), we can connect with our senses, more and more, to an infinitely more and more detailed scale, directly. It is not trivial in our hyper-stimulated world. Ultimately, the journey is personal. Albeit, as social animals, complicity from friends and groups can enrich and motivate us on our journey. But beware of select clubs for members only, beware of tribalism, beware of cults, beware of self-proclaimed holders of the truth... Why can't I end with a positive note?

The ability even the urge to ask, "what is the meaning of life", is incidental to the evolution of the neural circuits that generate predictive mental models, the very ones introduced by Kenneth Craik in his 1943 book "The Nature of Explanation". As Sam Harris said, we must live the question. We must live with the many revving thoughts that have nothing to do with the two evolutionary drives that incidentally led to millions of species, one among them being *Homo sapiens*: survival and passing of genes. However, it is also our capacity to think that allows us to learn, to create, again for the pure pleasure of it, but also to improve our quality of life, and in some cases to destroy us.

Etymology of the word "feeling"

The meaning of the word feeling drifted from its origin as signifying "sensation of touch" to define thoughts or emotions. I propose that examining the etymology of what it means "to feel" will allow us appreciating the title of this chapter, which is also the title of the book: "Feeling the Meaning of Life". Just as an intellectual exercise, I would like to make a parallel with the evolution of senses: from transducing and acting to the interposition of cells between primary senses and effectors, until the in between becomes "feeling" without the need of input and output. This is how the third chapter started, with the obscure sentence: "Between sensors and effectors, the in-between becomes sense, trying to make sense, compelled to make sense. Let go, breathe, and come back to your senses."

In the late 12th century, "to feel" refers to the physical act of touching, and to the sense of touch itself. Mid 14th century sees its first use as signifying "a conscious emotion" or an opinion (what one feels about something).

Most contemporary dictionaries concur that "to feel" has several meanings:

1. To use the sense of touch.
2. To become aware of through the skin; to use the sense of touch on.
3. To find one's way (literally or figuratively) by touching or using cautious movements.
4. To receive information by touch or by any neurons other than those responsible for sight, smell, taste, or hearing.
5. To search by sense of touch.
6. To sense or think emotionally or judgmentally.
7. To experience an emotion or other mental state.
8. To think, believe, or have an impression.

9. To sympathise; to have the sensibilities moved or affected.

Which one of these nine meanings best applies when trying to act upon the peculiar proposition "feeling the meaning of life"? How does one "feel" the meaning of life? Perhaps most definitions apply. Feel free to try each one by one; perhaps some definition will touch you more than others when thinking of the implied question: what is the meaning of life?

Have you tried applying each definition to the book title? If so, lets talk about them. A viscerally emotional component is felt when asking, "what is the meaning of life". Definitions 6 to 8 are tempting to use. You could also sympathise with the question itself (definition 9). What about the physical sense of touch?

It is my view that a way to live the question (as proposed by Sam Harris) is through definition number 2: To become aware of through the skin, to use the sense of touch on. However, I would also propose to modify definition number 4, to include other sensory modalities, particularly proprioception, which implicitly falls under the classification of touch in the 5 subjective categories of human senses.

One way of living the question "what is the meaning of life" is to stop everything and pay attention to your senses, while accepting the random flow of thoughts without acting on them, but rather by returning your attention to your senses. The most common of the senses felt in stillness and silence is your breath.

If we are going with definition 7, how are you feeling today? We are inquiring about your mood. Just as an example, there are many key players in threat signals and reward systems, all of which are key to determining what is our mood (Liu and Chaudhury, 2017). You might recognize some players: the amygdala (Amyg); the medial prefrontal cortex (mPFC); the hippocampus (Hipp), the locus coeruleus (LC), and the lateral hypothalamus (LH). The ventral tegmental area (VTA) is the hub for reward; this is where most dopamine is produced.

Figure 28

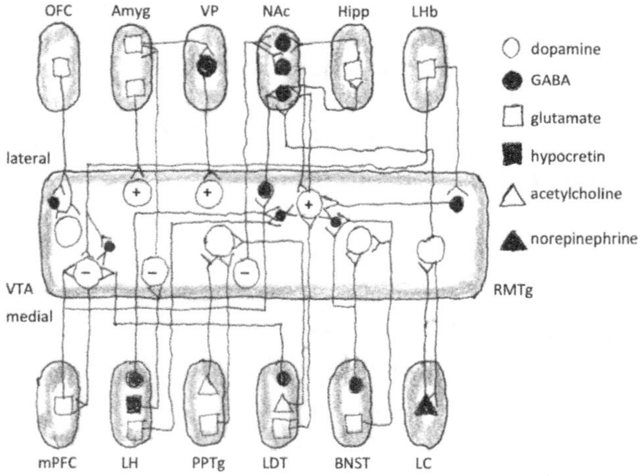

You can feel the air passing through your nostrils as well as your chest inflating and deflating. As you pay attention (which is synonymous to observing or meditating) to your breath, you let go, and with practice (repeating everyday for no particular reason) your breathing becomes your anchor, your beacon, and your source of bliss. What happens?

I would like to suggest that when you are in a state of silence and stillness, you let go, you make that leap of faith, you are no longer victim of the loop of thinking about thinking (Frith's meta-consciousness). Thinking about thinking can be a curse; another name for it: "to ruminate". When you ruminate, you drift endlessly, out of the physical reality your ancestors have evolved in. The expression "come back to your senses" is full of wisdom. You can connect to your senses, those arising from breathing: the rhythmic passage of air in and out of your nostrils or the rhythmic movement of your chest inflating and deflating.

Now, here is an interesting question: what is the origin of your rhythmic breathing? Where does it come from? You certainly can control how you breathe, but you can also let go to it, you can take that leap of faith. Try it now if you can, try to feel your breath without actually forcing anything, just letting go, trust yourself, do not make any efforts to breath, just let it be, even if you stop, trust me, you will eventually breathe again, just let go. As you do so, your breathing is on automatic pilot, it is controlled by your brainstem, more specifically, by assemblies of interneurons forming neural networks called CPGs; this is your "reptilian brain" taking care of everything. In silence and stillness, paying attention to your breathing, your brain devotes its full awareness to the outcome of CPGs. During this moment, there are no threats, no worries, no urge to answer questions, you are truly free, and thus can feel bliss. While stimulating the reward system leads to pleasure, the experience is ephemeral, leading to wanting more and more and more. The feeling of wholeness during repeated mindful breathing practice, in silence and stillness, tends to percolate outside of formal sessions.

We mentioned CPGs in the third chapter, when talking about the different levels linking sensors and effectors. Together, the group of interneurons forming the CPGs have the intrinsic ability to generate rhythmic patterned activity, with or without sensory inputs. From an evolutionary perspective, they are the first examples of interneurons able to generate activity independently from sensory feedback or any other patterned input. The evolution of CPGs made muscle coordination possible, allowing organisms such as lampreys (used as a model by Sten Grillner) to swim in their environment (Keijzer et al. 2013). As evolution followed its course, CPGs became involved in other rhythmic behaviours including crawling on earth, walking, feeding (chewing, swallowing) and breathing.

Briefly, in human, breathing is coordinated in the brainstem by distinct assemblies of interneurons that are respectively specialized in driving distinct aspects of breathing. These groups are under the control of the preBötzinger Complex, which coordinates all phases of the breathing cycle (Molkov et al., 2014; figure 29).

Of interest is that there is accumulating evidence of a reciprocal relationship between the preBötzinger Complex and brain regions associated with emotions. Christina Zelano and colleagues (2016) showed that natural breathing through the nose was associated with synchronized electrical activity in the limbic system (responsible for emotions), including the amygdala (fear detection) and hippocampus (memory). They also obtained behavioural results supporting that the breathing phase enhanced the discrimination between real and false threats as well as improved memory retrieval. In addition, Zamoscik and colleagues (2020) reported that mindfulness practice reduced the respiration pattern variability associated in some cases of depression.

Figure 29

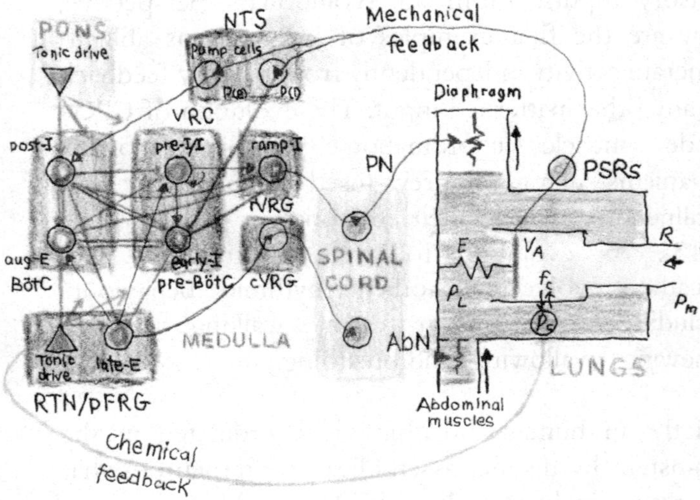

They previously had established that respiration pattern variability was linked with increased connectivity between the parahippocampal gyri and the posterior cingulate cortex, which tends to be more pronounced in states of depression (Zamoscik et al., 2018). Finally, Andressa Magalhaes and colleagues (2018) did a systematic review of 11 independent studies, and concluded that meditation was associated with increased activity in the two following brain areas: 1) prefrontal/frontal cortex, in line with increased levels of cognition and attention; and 2) insular cortex, in line with greater bodily awareness and potentially healthier emotional states.

The actual mechanisms underlying how the action of breathing itself might modulate cognition and emotion still elude us. However, promising and pertinent anatomical pathways have been proposed. Specifically, two distinct types of preBötzinger Complex neurons (DBX1 and CDH9+) project to the locus coeruleus, the main region where noradrenaline is produced and represents a hub for emotional control (Yackle et al., 2017). One digression, did you know that the uniquely high concentration of noradrenaline gives this nucleus a tint of blue? Locus coeruleus is Latin for "the blue location". This is only one of many other candidate pathways linking breathing CPGs and the higher brain centres involved in emotions.

Just to make a point that there does not have to be anything mysterious to how one might reach a state of bliss, here is a quote from an exhaustive review by Del Negro and colleagues (2018): "*PreBötzinger Complex projections to the suprapontine nuclei via direct or indirect pathways through the parabrachial nuclei or locus coeruleus or via olfactory nuclei, which receive respiratory-modulated input; vagal afferent input activated by large inspiratory efforts (note that vagal stimulation is effective in treating depression); afferent pathways activated by changes in blood gases; and collaterals of descending inputs that drive volitional or emotional breathing. The bottom line is that there are numerous connections between the brainstem breathing CPG and more rostral brain structures through which physiological, volitional and emotional stimuli that affect breathing in turn affect arousal, emotion and cognition.*"

The point I would like to make is that one cannot disregard these observations. It takes efforts to understand, and everybody has her-his own limits, including me as a sensory physiologist. These details are outside of my comfort zone. I indulge in quoting from this review to support the importance of facts. You do not need to know the details of how cars work (let alone their increasingly electronically controlled systems) to accept that there is nothing mysterious, and that in fact the functioning of car engines can be explained. In no way am I comparing cars with us, do not take me wrong here. Again I need to stress this:

Trying to understand the functioning of the brain does not mean to desacralize life and the world we live in. One must find the sacred in the material world, without ever being materialistic, more than ever. Observing with compassion is part of it.

Let me push this further as an exercise. I would like to propose that the source of the feeling of bliss through meditation, prayer, contemplation, involves the pre-Bötzinger complex interacting in a reciprocal manner with areas in the brain that deal with threat detection and emotions. Imaging studies have identified distinct regions of the brain that are activated when a state of bliss or enlightenment is reached, through religious practice or not (Newberg and Waldman 2016; Beauregard and O'Leary, 2007). While some researchers propose that these regions prove the existence of something greater than us, something outside the material world, I see here a link between the pre-Bötzinger complex and these regions. A common factor in reaching these states of enlightenment is to regularly impose oneself silence and stillness. Silence and stillness can be achieved through mindfulness, Zen practices, prayer, or contemplation, to name a few.

Ultimately one connects to a feeling of inner peace through letting go to your intrinsic rhythmic breathing, whether you are aware of it or not. That is, there are various ways to connect with your breath, consciously or not. Mindful breathing is not the only way. If you let go to your conviction that there does not need to be any meaning to life, or that you do not know, or that the meaning is God, Gott, Gud, Jesus, Heiland, Christ, Messiah, Yahweh, Allah, Shiva, Muhammad, Hu (Huwa), Hayyi Rabbi, the Buddha, Bhagavan, Bahà, Akal Murat, Akal Purakh, Xwedê, Jehovah, or any other external agent, if you let go truly and honestly in stillness and silence, you will connect to your breath in stillness and full awareness, knowingly or not. This state of wakeful rest is accompanied by what is known as "default mode network". At this very moment, specific regions of your brain will be highly active, much more so than when performing any focused tasks (Raichle, 2015). We talked about some of these regions; they tend to be located in the prefrontal and temporal lobes and the posterior cingulate cortex. The key is to practice again and again, with humility and sincerity, opening your heart. The outcome, I propose, is the same: inner peace. Letting go is counter-intuitive, hence the need to practice. There is recent evidence that activation of the default mode network via meditation might contribute to healthy aging (Ramírez-Barrantes et al., 2019).

In conclusion, how do we feel the meaning of life? We can connect to our breathing, and let go, setting time aside on a daily basis. At least, it is one way. How can we achieve this?

1- By being curious and carefully observing how our physical body has evolved and how it works. Without knowing the details, one can be open and accept scientific observations, until better ones, if it is the case, come forward and replace previous ones. This approach of educating oneself is compliant with leading a spiritual and sacred life. Reductionism does not imply to desacralize.

2- By stopping and connecting with our senses, such as epitomized by our rhythmic breathing. Frankly, in our modern society, meditation or mindfulness or contemplative prayer are often seen as inconveniences or even eccentricities. You must exert your right to be you, mostly because you are not infringing on anybody's liberty of expression.

Which brings me to what I consider to be the two major threats in our society: ignorance and convenience. How can one address these concerns in a humane fashion?

Chapter VII

Ignorance and Convenience

"*Nothing is more terrible than to see ignorance in action*"
Johann Wolfgang von Goethe

Ignorance means that you do not know certain things. We are all, therefore, ignorant in some respect. Here again are the words of Mary Catherine Bateson, quoted in the prologue: "*We are not what we know but what we are willing to learn.*" But there are those who will deliberately ignore facts that do not fit with their model of reality, those who purposely ignore well-established accessible facts, those who are not willing to learn. There is a term to describe this unfortunately pervasive behaviour (seen for instance with creationists, vaccine opponents, climate change deniers, homophobes, racists, conspiracy theorists, ufologists): wilful ignorance. It describes the denial of any sensory input that does not fit with one's inner model of reality. The underlying culprits are cognitive biases manifested in several forms, such as confirmation bias and causal bias. In the day-to-day English language, the use of the word ignorant has switched to imply deliberate ignorance. That is, intentionally "ignoring" established facts, evidence and/or reasonable opinions if they are contrary to ones belief. This can be achieved by inventing excuses such as: this source is unreliable or simply "false news", this experiment is flawed, or this opinion is biased. "Reasoning" becomes a circular argument where one cannot agree with a specific source because it is untrustworthy and because it is incompatible with his-her stance. Maintaining this denial of reality can entail outright refusal to read, hear or study the actual factual arguments.

"*If you think education is expensive, try ignorance*". Derek Bok

The cost of ignorance is insidiously permeating into multiple facets of our society. For instance, conspiracy theorists are rarely involved in constructive political engagements (Jolley and Douglas 2014); they instead lean toward political violence (Bilewicz et al., 2013) and tend to exhibit racist behaviours (Swami, 2012). As is the case with believers in alternative medicine, conspiracy theorists tend to be more at risk for serious health conditions such as HIV (Ford et al., 2013), to subscribe to the discredited findings that vaccines can cause autism (Jones et al., 2010) and to reject well-established medical treatments such as antipsychotics (Utinans and Ancane 2014).

"*It never ceases to surprise me at the infinite capacity of the human mind to resist the introduction of useful knowledge.*" Thomas Raynesford Lounsbury

By reading this book, you the reader are curious, open minded and starving for facts, for observations, for links, for context. However, nobody is immune to be biased, to not accurately take in the reality we live in. In our daily lives, we all unconsciously use mental shortcuts, known as "cognitive biases", which have been formed through: how we were raised; our previous experience; our cultural background; our education; our circle of friends, etc. While these shortcuts (such as the ability to read the mixed letters and numbers in the prologue) are convenient in our society, there is a cost we pay: our inability to accurately grasp the world inside and around us.

At birth until about 3 months, we do not have a fovea, and the image of a face consists of separate elements (such as mouth, nose, eyes) that do not form an actual face yet. Each element has features such as a smile. Now, when we read, the images formed, at least for some of us, might only capture features but not the whole picture. When asked to actually imagine a specific object mentioned in a book such as a house, it is hard to draw it as a whole, but before needing to draw it, the house did have salient features. Out of convenience, we might hold on to features rather than a whole complete picture.

Moreover, the images we recall from reading, tend to be vague because our visual memories are made of fragments and therefore are not like recalled photos. When reading words (such as with the text illusion presented in the prologue) we grasp whole words without taking into account all the letters, some of which are in fact numbers. We can, and do, live without complete pictures. Peter Mendelsund, in his book "What we See When we Read" takes us on a fascinating journey where we realize that everybody creates his-her own world from reading. Moreover, this world is an illusion since it has to be rebuilt every time we try recalling what we read. Commenting on reading stories, Mendelsund states: "*And the more we are immersed, the less we are able to bear upon the experience in which we are absorbed. Thus, when we discuss the feeling of reading we are really talking about the memory of having read*". Astute authors deliberately leave out certain details about characters, objects, events, so that we have more freedom to create personal images. In a way, being deliberately vague is an old tool used by leaders, politicians.

We will see below an example of a cult leader who deliberately stays silent, using his eyes to connect with his followers, and when speaking, banks on vague statements. This technique is infamous within fortune-tellers and tarot readers.

How seemingly paradoxical is it to address a chapter on ignorance and convenience using reading as an example. Writing and reading as well as speaking and listening, are our ultimate tools to pass on knowledge, to discuss, to grow, to attempt capturing the complex reality we live in. If anything, reading or listening to audiobooks or conferences, is essential to counteract ignorance and convenience. My point is that one must be cognisant of the traps we can fall into, whether it is through text or oral discourses. Another powerful mode of learning is via arts: paintings, sculptures, theatre, cinema, dancing, music, and any other forms of performing arts.

Ultimately, we construct our own images, our own reality, even when reading or listening. More and more we also rely on videos to communicate, which can add other dimensions to potential biases, images can be imposed, or also suggested. Regardless of how you take in the world, you are constantly vulnerable to cognitive biases. Numerous factors will affect how you behave, what decisions you make, what and whom you believe, how you behave and interact socially, and finally how and what you remember. There are hundreds of distinct cognitive biases, feel free to use your favourite search engine to look up

"List of cognitive biases". You will find an extensive repository for many such biases in political figures.

The two main ones underlying ignorance and convenience, according to peer-reviewed scientific papers, are "confirmation bias" and "causal bias". The former refers to "cherry picking" the facts that support your beliefs and dismissing others that go against them, as well as interpreting ambiguous evidence as corroborative, and focusing on select possibilities, all of which do confirm your beliefs. Such behaviour is reinforced by emotional stances and their desired outcome. The second bias, causal bias, is related to the fact that humans have evolved to detect and establish causal relationships as a mean to predict future events and adjust their behaviour accordingly. In the fifth chapter we mentioned the theory first established by Kenneth Craik about mental models as predictors of potential future outcomes to allow the implementation of the most adaptative response. Errors can arise, as we saw with examples of optical illusions in the first chapter.

Magicians bank on these flaws, where they purposely lead us to establish causal links that do not actually exist. There is increasing evidence that false attributions of cause and effects are in part responsible for why people believe in superstitions and pseudoscience (Torres et al., 2020). One important take home message is that the correlation of two events does not necessarily imply that one causes the other. How do we minimize causal biases? Teaching how to think scientifically, if you prefer another word than science, teaching how to think critically. What is factual and what is not. What are the actual underlying causes of the facts I am observing?

"*How do you progress as a culture if you set out to destroy any common agreement as to what constitutes a fact? You can't have conversations. You can't have debates. You can't come to conclusions. At the same time, calling out the transgressor has a way of giving more oxygen to the lie. Now it's a news story, and the lie is being mentioned not just in some website that publishes unattributable gossip but in every reputable newspaper in the country.*" Andrew G McCabe, The Threat, p. 255.

"*The more we teach people to simply accept anecdotal stories, hearsay, cherry-picked data and frankly, out-and-out lies, the harder it gets for people to think clearly. If you cannot think clearly, you cannot function as a human being. I cannot stress this enough. Uncritical thinking is tearing this world to pieces.*" Phil Plait, Bad Astronomy

How so very painful can it be for some to have to explain their political views in details with logic? Some fall back on emotions. Demagogic leaders bank on vague statements full of emotional black and white shortcuts that are not based in the complex reality we live in. The tools are remarkably similar between such dangerous leaders: scapegoating; fearmongering; lying; emotional oratory and personal charisma; accusing opponents of weakness and disloyalty; promising the impossible; violence and physical intimidation; personal insults and ridicule; vulgarity and outrageous behaviour; folksy posturing; gross oversimplification; and attacking the news media. The 45th American President comes to mind. Who has time for deep reflections? We are bombarded by stimuli. We are tired. We often mock someone who leans towards intellectual lingo or reflection. We are starving for light talk, and we feed on entertainment such as light TV shows and dubious YouTube channels. We are so very tired. We need to relax. How am I going to get people to think with this book? Who will read it? There is even the idea that this book is for an elite. I sincerely hope not. This book is for the curious, and the curious will share with others; perhaps teachers will read it. Perhaps its content will be shared on mainstream media. One can dream. The need is here.

"*I worry that, especially as the Millennium edges nearer, pseudoscience and superstition will seem year by year more tempting, the siren song of unreason more sonorous and attractive. Where have we heard it before? Whenever our ethnic or national prejudices are aroused, in times of scarcity, during challenges to national self-esteem or nerve, when we agonize about our diminished cosmic place and purpose, or when fanaticism is bubbling up around us - then, habits of thought familiar from ages past reach for the controls.*" Carl Sagan

Pseudoscientific beliefs tap in common cognitive biases, as we mentioned, the two main ones being confirmation bias and causal bias. It requires quite an effort to understand the documented events and mechanisms responsible for our evolution. It is much more convenient to believe that life was created by a supernatural being. Still today, despite the overwhelming scientific evidence to the contrary, creationism and intelligent design are thriving. There is also the notion of comfort in one's belief. As Michel de Montaigne aptly stated: "*Ignorance is the softest pillow on which a man can rest his head*". A few centuries later, Alexander Solzhenitzyn concurred: "*We do not err because truth is difficult to see. It is visible at a glance. We err because this is more comfortable*".

I feel a need to reiterate the quote from Immanuel Kant (from the reader's right): "*Laziness and cowardice are the reasons why such a large part of mankind gladly remain minors all their lives, long after nature has freed them from external guidance. They are the reasons why it is so easy for others to set themselves up as guardians. It is so comfortable to be a minor*".

Here is another pertinent reflection from a discussion I had with a previous student of mine, Dr. Greg Gilmour. Greg reflected: "*I was thinking that the (current) rise of populism coincides with the facts of everyday existence becoming too complicated. Resentment and outright rejection of expertise of any kind is an alternative to accepting a lack of understanding and control. Substitute a version of reality that's simple, digestible and whose palatability isn't constrained by facts.*"

While one wants to open a constructive discussion with believers, those who rely on pseudoscience are descending a slippery slope toward magical thinking, using oversimplification, confirmation biases and causal biases. Pseudoscientists are at times fuelled by attraction to fame and money (speakers, gurus, solutions, wow factors); they prey on sincere spiritual seekers; and they claim to hold the ultimate answers to natural questions: what is the meaning of life, what is my purpose in life, why are we here, what is the ultimate truth? I say: follow the money and the power that these leaders have over their followers. In his book "Waking up: A Guide to Spirituality Without Religion" Sam Harris warns us about predatory gurus, fanatic preachers, about the "holders of truth". Lets hear Bertrand Russell again (yes I have a bias toward his thinking process): "*The whole problem with the world is that fools and fanatics are always so certain of themselves, and wiser people so full of doubts.*" Doubting everything is after all a key process of the scientific approach.

Why do some follow, often blindly, those who claim to hold the truth? They can be individuals, groups, political parties, sects, religions etc. Another sceptic, Michael Shermer in "The Believing Brain" eloquently reviews the neurological basis of beliefs. It is beyond the scope of this book to go over the exhaustive scientific literature describing the neurophysiological changes underlying beliefs. Allow me to present one important player, the neurotransmitter dopamine. Dopamine plays a key role in the reward system and in learning by associations. For instance, Skinner trained rats to associate a specific stimulus, such as a sound, with a reward. After training, the sound alone was sufficient to increase dopamine levels. He called repetition-induced behaviours such this one: "operant conditioning". In human, Christine Mohr and Peter Brugger established a positive correlation between brain dopamine levels and beliefs in ghosts, gods, spirits and conspiracy theories (Krummenacher et al., 2010). Higher dopamine levels were linked with the tendency to extract meaning out of noise, to make sense out of nonsense.

Back to the letters and numbers in the prologue, your ability to make sense of the scrambled letters and numbers could be related to higher dopamine levels. Presented at the very beginning of the book, you might not have had time to be bored, and plus, you saw it as a challenge and a reward to solve a puzzle. In brief, believers have a tendency to find patterns where there is none, to find meaning when there is none, to read too much into a specific situation, especially if this situation is repeated again and again.

I cannot help but share my own experience with a holder of the truth, John de Ruiter, who happens to profess in the same city where I reside, Edmonton, Canada. The founder of the College of Integrated Philosophy, "John" (as his followers call him), started as a Christian preacher, espousing Jesus' look, when at some point he claimed he held the truth; there was no more need for the bible. This led de Ruiter to upgrade his bookstore meetings to a large auditorium venue, the Oasis Centre, which he had built for his teachings but also to be rented (at a cost) for weddings and various events.

During his three weekly meetings (350 attendees on average, each paying ten dollars per meeting), de Ruiter essentially stares in silence at each member of the audience, one by one, for what feels like an eternity (up to three hours) during which his face is displayed on large screens. At some point "John" will invite an attendee (carefully pre-screened, who's face now appears on the screens) to sit in front of him on the elevated stage. Before each meeting, you have the opportunity of purchasing a meal and/or beverage so you can share the same space as "John; he might even talk to you, slowly, and yes, blink. More recently, his blinking has increased, which has inspired his followers to do the same, they blink often. One can also attend weeklong seminars for two hundred dollars.

Over the years, de Ruiter had his share of scandals, cheating on his wife with two sisters (Benita and Katrina von Sass) and even publicly admitting having had a "Calling" to justify sexual relations with many of his female followers (Horsley, 2017). According to a

Globe and Mail article by Jana G Pruden (published on November 25, 2017), Benita von Sass wrote an affidavit in 2009, in which she estimated de Ruiter's personal assets at nearly $9-million, including his equity in the Oasis Centre, a $75,000 monster truck, the home purchased by Katrina and put in his name, and a personal income estimated at $232,000 a year. I do know an ardent follower who moved to Edmonton so he could be with "John", not only at the Oasis but during his many worldwide trips, helping his spiritual leader cover travel expenses. Sadly, the follower's devotion reached a point at which his partner asked him to choose between her and "John". "John" won; my friend lost a wonderfully loving partner. For my friend and all other followers, "John" is the living embodiment of truth, he is the Messiah. I asked my friend: what will happen when "John" dies? He replied: so will the truth. There are rumours that "John" might be grooming his son. Why do his loyal followers not see through de Ruiter's obvious manipulative, predatory, and chiefly self-serving behaviour?

While vehemently refuting the title of cult leader, de Ruiter employs well-established patterns attributed to gurus. According to Paul Joossee (2006), it is all about de Ruiter's silences. Followers can project that the leader understands their deep questions. On the other side, followers can also feel dominated by their leader's silence, as if he were punishing them. Finally, the interminable staring, compounded by the leader's silence, could induce a feeling of intimacy typically reserved for lovers, in other words they might feel reciprocal love, and yes their dopamine levels will likely increase, among other changes in brain activity (Harris et al., 2008).

Not only charismatic truth bearers but also mostly well-intentioned advice givers, abound. It is as if the instruction manuals were more complicated than the task they aim at achieving. At least, this is the feeling I get when browsing through the self-help book section in a bookstore or a library or online. And why is it that each of these mentors have to publish several books, and in some cases, produce endless series of DVDs, or online subscriptions to teachings, and of course conferences, many conferences, all of which are available for a "moderate" fee? Pseudoscience is ubiquitous and yes: money is involved. Believers are willing to pay financially and otherwise. In the best-case scenario when seeking health benefits, believers might experience a placebo effect, which is real, but healing might need several reiterations at a dear financial cost (Nahin et al., 2009). In the book they edited "Pseudoscience: The Conspiracy Against Science", Allison and James Kaufman gathered scientists who use evidence based approaches to respond to pseudoscientists. Reading the copiously referenced 510 pages book, one feels the frustration of having to repeat facts, rebut the same erroneous arguments, inexorably putting scientists in a defensive position. How did scientists end up on the accused's bench? Their detractors are not bound to the same rules. Like the cavemen, detractors get dopamine surges by winning without using facts. They win by getting people to doubt science. This is where wilful ignorance makes our society sick as a metastatic cancer. I feel as if I had to repeat again the same arguments as if fighting cancer. I feel the frustration. I will let the outspoken advocate for evidence-based psychological treatments Scott Lilienfeld summarize valuable lessons in the Allison and James Kaufman's book (my own indulgence in convenience):

"1) We are all subject to cognitive biases; 2) We are largely unaware of our biases; 3) Science is a systematic set of safeguards against biases; 4) Scientific thinking doesn't come naturally to humans; 5) Scientific thinking is exasperatingly domain-specific. Even Nobel Prize winners can fall prey to pseudoscience in fields outside their area of expertise; 6) Pseudoscience and science lie on a spectrum; 7) Pseudoscience is characterized by a set of fallible, but useful, warning signs such as an absence of self-correction, overuse of ad hoc maneuvers to immunize claims from refutation, use of scientific-sounding but vacuous language, extraordinary claims in the absence of compelling evidence, over-reliance on anecdotal and testimonial assertions, avoidance of peer review, etc.; 8) Scientific claims can be wrong. Pseudoscientific claims differ from erroneous claims in that they are deceptive: they appear to be scientific, but they are not. 9) Scientific and pseudoscientific thinking are cut from the same basic psychological cloth. Heuristics (mental shortcuts or rules of thumb) are invaluable in everyday life, but when misapplied they can lead to mistaken conclusions; 10) Skepticism differs from cynicism. Skeptics must guard against dismissing implausible claims out of disconfirmation bias."

Just as a reminder of the societal cancer that represents pseudoscience, below is a list of current examples of fake approaches proposed to live a healthier life. Who does not want that? Not the sham aspect but the certainty of a better life. Get your wallet! Before I do so — not get my modest wallet but present the infamous list — I just had a thought about compassion. Remember when I stated that the point of beliefs such as religions was not to simply say, "There is no God" but rather to understand why people believe and to acknowledge that the ability and tendency to believe is an offshoot (exaptation) of the evolution of our brain to create mental models? Using compassion, one has to realize that beliefs in external agents such as God or Gods or other entities, can act as stabilizing beacons, with concrete measurable changes in brain activity. Here I am comparing pseudoscience with a societal cancer. Pseudoscientific beliefs can exert measurable effects, analogous to placebo effects (also called "sham" groups in experimental settings), as I mentioned above. The difference is that pseudo-therapies are implemented on a false premise and at a cost. I will provide examples considered as pseudoscience, in that they are not supported by scientific evidence. Just to complicate matter, nothing is black and white.

While science is the best tool we have to grasp reality, the scientific approach remains imperfect. As such, we shall be humble and not hastily judge those who choose not to trust science. You can examine the limits of science for each of the examples below. While you might categorize some examples as obviously flawed, others might fall into a greyer zone: Acupuncture and Acupressure; Adrenal fatigue; Alexander Technique; Alternative cancer treatments; Anthroposophical Medicine; Apitherapy; Applied Kinesiology; Aromatherapy; Astrology; Attachment therapy; Auriculotherapy; Autistic enterocolitis; Ayurveda; Balneotherapy; Bates method for better eyesight; Biological terrain assessment; Biodynamic Agriculture; Biorhythms; Blood type diet; Body memory; Bodynamic somatic psychology and analysis system; Bowen Technique; Brain gym; Candida hypersensitivity; Chakra healing; Chelation therapy; Chiropractic; Chromotherapy; Colloidal silver; Colon cleansing; Craniosacral therapy; Crystal healing; Cupping therapy; Dianetics; Detoxification; Ear candling; Electrohomeopathy; Electromagnetic hypersensitivity; Energy medicine; Facilitated communication; Faith healing; Fasting; Feng Shui; Feldenkrais method; Functional medicine; Germanic new medicine; Graphology; Hair analysis; Healing jewellery; Hemi-Sync; Hexagonal water; Homeopathy; Hypnosis and Hypnotherapy; Iridology; Koranic scientific foreknowledge; Leaky gut syndrome; Lightning process; Macrobiotic diets; Magnet therapy; Memetics; Morgellons; Moxibustion; Myers–Briggs type indicator; Nambudripad's allergy elimination techniques (NAET); Naturopathy; Negative air ionization therapy; Neuro-linguistic programming; Oil pulling; Orthomolecular medicine; Osteopathic manipulative medicine; Palmistry; Parapsychology; Primal therapy; Polarity Therapy; Psychokinesis; Pulse diagnosis; Quantum mysticism; Radionics; Reiki; Reflexology; Rolfing; Rumpology; Therapeutic touch; Tin foil hat; Traditional Chinese medicine (TCM); Tree healing; Urine therapy; Vitalism; and as supported by the 45th American president, Wind turbine syndrome.

Let us single out one item from the long list: psychoanalysis. Is it science or pseudoscience? While Freud's approach was revolutionary and remains pertinent today in the field of psychopathology, it received vibrant criticisms from well-established brain scientists including Ramón y Cajal and the philosopher of science Karl Popper. Popper had a brief honeymoon with psychoanalysis, followed by the critique that predictions from this approach are not amenable to experimental verification. More recently, other scientists have expressed reservations (to put it diplomatically) with psychoanalysis, including psychologist Steven Pinker, linguist Noam Chomsky, evolutionary biologist Stephen Jay Gould and physicist Richard Feynman. For what some might consider the nail in the coffin, one could read (if you have not done so yet) the 2017 book by English professor Frederick C. Crews, "Freud: The Making of an Illusion". A second quick example: let us consider homeopathy, which 34% of Europeans consider being “scientific” (European Commission, 2005).

At the time of writing this book, in 2020, in the midst of the COVID-19 pandemics, public-health providers in Canada and the United Kingdom are offering to cover the cost of homeopathy to tackle a retroviral infection. One would like to assume that public-health providers should lead by example and not rely on pseudoscience; however, their gesture promotes that a scientifically implausible remedy can work against a retrovirus, SARS-CoV-2 (severe acute respiratory syndrome coronavirus 2). Proponents of homeopathy rely on an external agent, quantum physics, to hold on to a scientific explanation. As we saw with dualists, quantum physics is an "all you can eat buffet" when it comes to relying on external agents. But there is more than homeopathy. In Canada, chiropractors, naturopaths, herbalists and holistic healers are advertising approaches to treat COVID-19. Ignorance is costly to society. How do you address misinformation? Rather than confronting believers directly, the pre-emptive dissemination of factual evidence is one option proposed by Gorman and Gorman (Scales et al., 2020). The rationale is to make people more resistant to false information. One must resist the temptation to ridicule those who believe in pseudoscience, but instead engage the curious and convince the unconvinced (Scales et al., 2020). Not all beliefs in pseudoscience should be considered on the same basis, the roots could be quite different.

Considering all subjects that fall under "pseudoscience", vocal advocates for alternate medicine might be shocked that their beliefs fall under the same category as perhaps more sensational or extreme examples. Take "Ancient Aliens": Fifteen documentary-style seasons presenting historical texts, archaeology, and legends as evidence of past human-extra-terrestrial contact. How about 23 seasons of "Most Haunted", a series showcasing paranormal investigations into haunted locations, using psychic mediums and scientific equipment, aired since 2002 and still on-going. Another example? "Ghost Hunters", 13 seasons aired since 2004, which follows a group of real-life paranormal researchers investigating haunted houses, starring trade plumbers Jason Hawes and Grant Wilson who lead "The Atlantic Paranormal Society". To cut to the chase, a quick Internet search retrieved 31 paranormal TV shows and "documentaries", many of which qualified as "true" and "scary". Here is your guaranteed entertainment. At the end of the day, series and "documentaries" are money makers. The cost is that some viewers buy into it (no pun intended).

In the U.S.A., repeated surveys confirm that the majority of its population acknowledges holding paranormal or pseudoscientific beliefs (Gallup and Newport 1991; Orenstein, 2002; Rice, 2003). Three out of four Americans believe in at least one type of paranormal activity (Stark, 2012) and half subscribe to conspiracy theories (Oliver and Wood, 2014). In the mid-eighties, in Canada and Great Britain, a link was established between believers and scientific illiteracy as defined by the inability to correctly identify basic scientific principles (McLeish, 1984). However, more recently, and conceivably of further concern, two American studies published in the Skeptic Magazine (non-peer reviewed), reported that the level of science education as a whole does not predict whether someone will be less inclined to believe in paranormal and pseudoscientific subjects (Walker et al., 2002; Ryan et al., 2004).

You will ask, what is the point of learning science if it does not make you less susceptible to fall for pseudoscientific beliefs? Part of the answer lays in a 2018 peer-reviewed study published by Biologist James A Wilson. The key is not to remember scientific facts only, but rather to develop critical thinking, and unfortunately, teaching establishments rarely include in their curricula such courses. Wilson studied the impact of a 3 credits 4th year University lecture-based course, "Science and Critical Thinking" on beliefs in pseudoscience. A total of 313 students filled pre- and post-course questionnaires. He found that: "*Following the critical thinking course, overall beliefs in paranormal and pseudoscientific subcategories lowered 6.8–28.9%, except for superstition, which did not significantly change. Change in belief had both a gender and religion effect with greater reductions among religious students and females.*" (Wilson, 2018).

Here is another example of a promising intervention, which one could say is part of critical thinking. It addresses causal biases in adolescents. Barbeira and colleagues (2013) taught a group of students how to scientifically establish a link between a cause and an effect: comparing the probability of an outcome in the presence or the absence of the potential cause. Here is how the study was designed: two groups were compared, one that received the teaching (intervention group) and the other that did not (control group). This is what I call preaching by example. Both groups performed a "... *standard contingency learning task that involved fake medicines that typically produce causal illusions. The participants in the experimental group made more realistic causal judgments than did those in the control group, which served as a baseline.*" (Barbeira et al., 2013). Where is the entertainment factor, why are there no reality shows about this? How do we reach into the deepest corners of the world? Call me idealistic, but this should be taught and passed on to all human beings. As much as we teach our kids to walk, to be toilet trained, to be socially appropriate, we must teach them how to think critically.

We need not only facts but also critical thinking in our society, more than ever. Not respecting facts and not thinking critically about facts — making biased and irrelevant links — are one of the most severe threats to our society; they are real threats. James A Wilson states: "*All of these behaviors represent a socially destabilizing force, and potentially a widespread health hazard. Counteracting unsupported beliefs, by teaching critical thinking and promoting science literacy, is therefore important to the stability and health of society.*" (Wilson, 2018). At the expense of an open-minded and so-called constructive pedagogic approach, pseudoscience finds its way even in science classrooms (Mugaloglu, 2014).

Well-intentioned teachers might underestimate how hard it can be to have rational discussions with students defending pseudoscientific or absolute beliefs. Instead of teaching critical thinking, the teacher becomes a social negotiator facing group pressure to be at least open to the possibility that pseudoscientific beliefs cannot be disproved with absolute certainty. The role model played by the teacher as a compassionate human being compounds the task to inculcate critical thinking. This is why I would advocate for the earliest implementation of critical thinking in schools and in our society in general. In an ideal world, parents or legal guardians should preach critical thinking by example. There is a societal component to beliefs in pseudoscience and the paranormal. For instance, compared to Americans, Scandinavian children have a low propensity to believe in the paranormal (Harraldsson and Houtkooper 1996; Tobacyk and Pirttilä-Backman, 1992).

Regardless of our cultural background, we have a natural inclination to reject new evidence because it goes against what we've always done, or believed for so long. Dr. Semmelweis, a Hungarian physician, thought it might be a good idea if doctors disinfected their hands between morning autopsies and delivering babies in the afternoon. After some resistance, doctors eventually did obey the strict orders to wash their hands with a disinfecting solution. The mothers' death rates, caused by puerperal fever, plummeted. This incident became so famous that it led to the name: Semmelweis reflex. Here are just a few examples: 1) stomach ulcers are caused by Helicobacter pylori provided from food, and not by stress or spicy food; 2) tastes are not mapped on the tongue; 3) sugar does not make kids hyperactive; 4) we use all our brain, 100%, but in about 2% bursts due to high energy demands; 5) leprosy is not transmitted by skin contact; 6) stretching before doing sports does not diminish the risk of injuries; 7) eating before swimming does not increase your risk of cramps. These are concrete examples, but how about fanatics preaching in the context of politics, religions, sects, and conspiracy theories?

How again can you compassionately have a conversation with someone who believes in pseudoscience? You could approach this person by saying: we can live the sacred, be enlightened, nourish ourselves, emancipate ourselves, all from the facts (current evidences) without having to resort to beliefs. How do you convince people that vaccines are good for kids, and that handguns are dangerous? Presenting hard facts does not work; believers will ignore them. One could appeal to their emotions, but this would go against the goal of promoting science. Regardless of your approach when discussing with believers, they might answer:

> *I hold beliefs that I know are true with absolute certainty, and it is with all my soul that I will defend them. Science is simply not sufficiently advanced yet to prove my beliefs, I have the proof, look at all those phenomena that we cannot yet explain, such as energy transferred after death, telepathy, and trips out of our body. I have my beliefs and you have yours. Vaccines cause autism and handguns keep my family safe. You cannot prove that I am wrong, because you have your own beliefs dear scientist.*

Troy Campbell (University of Oregon) and Justin Friesen (University of Winnipeg) study how one can hold to his-her-their beliefs. Their findings support that having false views can be classified as a disease, which one can correct with a logical perusal of facts and most importantly education that starts at an early age. Feeding facts into a discussion with believers can in principle lessen their biases. As we discussed, people can wilfully ignore facts. The two researchers remain optimistic when confronted with this real disease. As part of a child's upbringing (and this applies to adults as well), one must be encouraged to accept ambiguities, to think critically and to recognize and reject strict ideologies. Here is an example of their research: Campbell and Friesen studied the opinion of people with regard to whether same-sex marriage should be legalized (Campbell and Friesen, 2015). A factual rebuttal against would require facts such as: "*children raised by same-sex parents are worse off — or just as well-off — as children raised by opposite-sex parents*". But what if those facts contradicted one's views? They presented one group with made-up "facts" supporting their opinion and the other with also made-up "facts" that contradicted their stance. When faced with facts opposing their views (regardless of being pro or con) the participants resorted to the argument that the issue is not about facts but more about a moral conviction. Interestingly, when facts supported their opinion, they maintained the opposite view: facts matter more than morality.

Let us continue on the subject of homosexuality. I personally have been deeply saddened by the news of a student who died by suicide; he could not come out of the closet. His gesture was literal and profoundly perturbing: he took an overdose of pills in the closet of his bedroom, where his roommates found him dead two days later. His parents' fundamentalist religious beliefs were clear: Romans 1:26 says it is disgraceful and dishonouring. First Corinthians 6:9

says it will keep a person from the kingdom of God. First Corinthians 6:18 says that homosexual behaviour is a sin against one's own body. I heard two years later that his parents are now supporting people who want to come out. The cost the parents and their son both paid is unbearable.

One can state that because homosexuality does not allow passing DNA over generations, it is against nature. Is that it? Depending on when and where you lived, this might be it indeed. Lets get back to human vision, where fine details vision covers only 5% of our visual field, the brain filling in the rest with the illusion that we see all with details. Lets get back to the text with numbers and letters in the prologue; in addition, our brain relies on learned patterns, creating an image that fails to capture the actual image. So, depending on when and where one lived, this is the message they might have learned – homosexuality is against nature, it is wrong – well this is the 5% detailed vision and filling in without actually capturing what wider context is accessible, is right here, right now. Most sexual relations do not lead to the passing of DNA; some that do can involve hatred, ignorance, poverty of means, passing perhaps genes but not the culture of respect of all forms of life.

I shall let the writer Reinaldo Arenas have a say here: "*Leonardo da Vinci was homosexual, so was Michelangelo, Socrates, Shakespeare, and almost every other figure that has formed what we have come to understand as beauty.*"

I do not know about you, but I would like to understand more about why it seems so hard to grasp reality. Our brain can build highly subjective mental models, independently from any sensory inputs, which do not represent the requirements for survival and reproduction that allowed us to evolve as *Homo sapiens*. Part of the answer is that we are social animals. Cognitive scientists Hugo Mercier and Dan Sperber, in their book "The Enigma of Reason", articulate: "*Reason is an adaptation to the hypersocial niche humans have evolved for themselves*". What seems rational from an intellectual point of view takes a different meaning when "reasoned" from a social interactionist perspective.

Social circles are incubators for a rich array of cognitive biases, with the most common being confirmation bias, as we discussed above. Members of social circles feed each other with reciprocal confirmation. Internet social media are a haven for confirmation biases: algorithms are designed to post similarly minded opinions on your social media account. Mercier and Sperber did a study in which participants were given a chance to modify their response to specific questions when presented with contradicting answers from other participants. The trick here was that the answers presented to them as someone else's were actually their very own answers, and vice versa. The half portion of the group that was oblivious to the trick became the study subjects. Close to sixty per cent vehemently rejected their own initial responses, using highly critical arguments. According to Mercier and Sperber, such behaviour supports that reason evolved to prevent us from being banned by members of our group.

Members of closely-knit small tribes valued they social ranking, especially when their status prevented them from risking their lives on the hunt. In such social context, using reason did not matter as much as winning arguments, because winning could determine life or death. In their book "Denying to the Grave: Why We Ignore the Facts That Will Save Us", Jack and Sara Gorman argue that maladaptive attitudes, such as those displayed by vaccine deniers, might arise from this same originally adaptive trait of human evolution. The main culprit, confirmation bias, has a physiological component: dopamine surge from processing information supporting one's belief. The pleasure-reward centres are activated when one feels they are right, even though they are not. They still get their dopamine rush.

Based on findings that the first *Homo sapiens* tribes date around 100,000 years ago, the results of this study provide an example in which the environment changed too quickly for natural selection to lead to the most adaptative behaviour.

Steven Sloman and Philip Fernbach concur that original social behaviour in the earliest evolution of *Homo sapiens* has led to a maladaptation in our current society. Winning an argument rather than struggling to be factually accurate, is how we have evolved as human, at least it is a hypothesis to further study. In their book "The Knowledge Illusion: Why We Never Think Alone", Sloman and Fernbach recount asking Yale graduate students to rate their understanding of everyday devices, including toilets, zippers, and cylinder locks. While confident they could actually explain the details, most Ivy League students failed miserably to do so. The authors referred to this behaviour as the "illusion of explanatory depth". Such illusion is analogous to the commonly observed inability of political partisans to exhaustively articulate their views in details with logic, as mentioned above.

As an avid cyclist, I would assume I could easily draw a bicycle. It turns out that drawing the frame itself is simply not as intuitive as you would assume. Try it. We tend to believe that we know more than we actually do, and the culprit is that tasks are extensively divided in our society.

Dating from the first hunting tribes, humans have relied on each other's expertise. Specialization was further exacerbated to meet the needs that sustained sedentarisation. Basic needs were agriculture and the domestication of animals, both of which were essential to feed larger tribes, and eventually complex civilizations with numerous specializations. New needs emerged, such as storing food and keeping records, which led to the invention of writing and mathematics. Attributing values facilitated the distribution of goods: money was invented. As civilizations developed more advanced technology, convenience became a goal in itself. How can we make our life easier? How can we produce more with less effort? How can one make more money? How can I gain a higher social position to "optimize my survival"? Such environmental, technological and societal changes provide a fertile ground for innate behaviours to become maladaptive. Progressively, humans changed their environment to accommodate endless needs, some driven by selfish goals. New needs arose from the changing environment itself: an uncontrolled loop, where needs create more needs. Moreover, individual members of the group became unaware of the global picture. As convenience and a lackadaisical attitude dominated, individual ignorance of the world became more prevalent.

In conclusion, some of the behaviours that have allowed us to evolve as *Homo sapiens* are no longer adaptative in our current environment. The question is: how do we make sense in the environment we are rapidly transforming? The short answer is we need to educate ourselves; we need to reflect on what is happening, on what are the consequences of our actions. The hamster must stop running in its wheel and take in the world around him. When over 99% of the world's population has to work long hours at a minuscule salary that is barely sufficient to survive, how can they be motivated to stop and think outside of their own routine?

"*There is more to us than we know. If we can be made to see it, perhaps for the rest of our lives we will be unwilling to settle for less.*" Kurt Hahn

Chapter VIII

Making Sense in our Changing Environment

The oldest fossil records of *Homo sapiens* where found in Jebel Irhoud, Moroco; they date 300,000 years (Hublin et al., 2017). In only the last iota of its existence as distinct species – starting 3,500 years ago, with the Mesopotamian civilization – humans changed their physical and social environments. The most drastic transformations correspond to the dawning of the industrial revolution, 250 years ago, culminating to the ubiquitous use of the Internet, social media and smartphones over the past 20 years. The outcome is that certain of the then adaptive traits have become recently rapidly maladaptive. In the previous chapter, we just touched upon the hypothesis that wining an argument might have been more important than using logical explanations based on hard facts. Here, I would like to expand on how the maladaptation of our senses in the world we are altering, is affecting us.

You might ask: what is the link with the meaning of life? My answer (not that I want to win any arguments) is: we can make sense of life by connecting with our senses and taking in what is inside and around us. We have so many reasons to live a rich and meaningful life. There is so much to do in order to live fully and in harmony with nature. Honestly, we can do much better as a society to take care of all life forms and our environment. We too often behave as hamsters running in a cage. Let us truly feel what is happening and act accordingly. The Hopi word Koyaanisqatsi, meaning "life out of balance", was quite visionary for tribes formed around year 1,100 to 1,300 in America. It is likely that the word might have been inspired by the arrival of Spanish occupiers in 1540s, who enslaved Hopi people to produce more crops and goods, culminating to the Pueblo Revolt of 1680. Watching the 1982 film produced and directed by Godfrey Reggio, aptly named "Koyaanisqatsi", is an eye opener. Unfortunately, activists warning us that we are out of balance are too often marginalized, even often repressed, in our society.

We need to come back to our senses.

Smelling:

"*Smell is a potent wizard that transports you across thousands of miles and all the years you have lived. The odors of fruits waft me to my southern home, to my childhood frolics in the peach orchard. Other odors, instantaneous and fleeting, cause my heart to dilate joyously or contract with remembered grief. Even as I think of smells, my nose is full of scents that start awake sweet memories of summers gone and ripening fields far away.*" Hellen Keller

Lets start with the most obvious change in our environment, air pollution. Air pollution pauses the most severe environmental threat to humans and in general to all forms of life on earth. In fact, air pollution represents the main cause of human illnesses. Breathing through your nose, you would expect deleterious effect on the sense of smell, and you would be right. Alaska University professor Kara

Hoover, studies the maladaptation of our sense of smell. She says: "*Our sense of smell evolved in very odour-rich landscapes in which we were interacting regularly with the environment*", adding that "*Now today we're not interacting with the environment and we're in very polluted places.*" (quoted in Johnston, 2017). The key issue in studying how pollution affects the sense of smell is that olfactory impairments do occur naturally with aging and accompany neurodegenerative diseases (Lafreniere and Mann, 2009), autism spectrum disorders (Bagasra et al., 2018), brain injury (Callahan and Hinkebein, 2002), certain medications (Jafek et al., 2004), and exposure to pesticides (Shrestha et al., 2019). The question is whether olfactory issues are causes or consequences. In most of the cases listed above, they are likely symptoms, or could there be exceptions? Chronic exposure to air pollution has been linked with neurological and psychiatric disorders such as cognitive decline, dementia (including Alzheimer's disease), anxiety, depression, and suicide (Thompson, 2019). In laboratory rodent models, ozone and particulate matter have been shown to activate the hypothalamic-pituitary-adrenal axis and to release stress hormones (corticosterone, the rodent equivalent of cortisol in humans) as part of the neuroendocrine stress response. Chronic stress response itself can induce severe biochemical and structural alterations in the brain, which would explain impairments in cognition and mental health. What complicates the matter is that chronic activation of the stress response by air pollutants is typically compounded by other variables such as noise pollution. As we will see with other senses, there are many other aspects of human-induced changes in our environment that also impact on our senses and on our overall physiology, including our brains; one that comes to mind is cell phones, but lets move on with hearing first.

Hearing:

"*Soon silence will have passed into legend. Man has turned his back on silence. Day after day he invents machines and devices that increase noise and distract humanity from the essence of life, contemplation, meditation...tooting, howling, screeching, booming, crashing, whistling, grinding, and trilling bolster his ego. His anxiety subsides. His inhuman void spreads monstrously like a grey vegetation.*" Jean Arp

We have not evolved to be submitted to constant noise. As hunters, our ancestors would move stealthily in the forests, listening to the mildest of sounds epitomized by the rustling of leaves. Threats would be signalled by sudden and brief loud noises such as a the roaring of a protective mother bear, the snarling of an inquisitive lion, the bellowing of a rhinoceros, the trumpeting of charging elephants, or the thundering dark sky. A recent study by the World Health Organization established that after air pollution, traffic noise comes second as one of the worst environmental stressors for humans (Münzel et al., 2020). Over four billion people live in densely populated cities (Ritchie and Roser, 2018) where traffic noise predominates almost 24 hours per day.

Urban settings are a very recent phenomenon in human history, let alone in the evolution of *Homo sapiens* from the first hominine ancestor (*Sahelanthropus tchadensis*), 6.5 millions years ago. In addition to impairing our hearing, chronic noise exposure affects our threat response system; this means that our sympathetic nervous system is constantly activated:

cortisol levels remain elevated, which triggers inflammatory and oxidative stress pathways, inducing vascular and neuronal dysfunction (Hahad et al 2019). Another effect of persistent noise is that it precludes restful deeper stages of sleep and prevents our blood pressure to rise and fall, as it should along the day and night cycle (known as the circadian rhythm). Hence, as a consequence of sleep disturbances, chronic noise exposure has been linked with vascular defects encompassing hypertension, vasoconstriction, and other cardiovascular adverse effects such as myocardial infarction and strokes (Dzhambov and Dimitrova, 2016; Khosravipour and Khanlari, 2020; Münzel et al., 2020). The list of harmful consequences is actually much longer, but statistical correlations vary between studies: weight gain (Wang et al., 2020); personality changes and violent reactions (Elizondo-Garza, 1999), depression and anxiety (Dzhambov and Lercher 2019), reduced self-care and tendency for social isolation (Hammer et al., 2014).

In her book "Sound and Noise: A Listener's Guide to Everyday Life", Marcia Jenneth Epstein reminds us that we adapt to any types of noise by ignoring it. What she considers as noise here encompasses a vast range of sources, from traffic, ventilation, heating or cooling systems, even the lies of politicians. We burry noise under background music using our smartphone, by attending loud concerts, by riding with noisy engines and steroid boosted sound systems. Some just like to pump up the jam. "*Because we grow up with a background of mechanical roars and crashes as well as the soundtracks of broadcast urgency, we learn that they are* "normal". *Normal does not mean entirely safe, however. Our refined biological alarm systems can be in constant overdrive as a result of ambient urban and mechanical soundscapes as they contribute to a host of stress-related insults to health: anxiety, muscle tension, headaches, insomnia, irritability, fatigue. Like the noise, these conditions have become normal, especially in racialized and impoverished urban districts. Thus, awareness of sound carries an element of social justice, an imperative to provide accessible havens of quiet in neighbourhoods that are also in need of affordable housing and nourishing food.*"

Marcia Jenneth Epstein also reminds us that some types of sounds can be soothing, even heal, such as the natural sounds of the rustle of leaves, waves, bird songs, crickets, frogs, etc. Listening to such natural sounds mindfully with loving-kindness is also a way to "coming back to our senses". We can also truly listen to each other with compassion.

Seeing:

Chronic exposure to air pollution directly impacts on eye health, as does water pollution. Deleterious effects on the eyes range from minor irritation to retinal bleeding. "*Taking a shower in water contaminated with chemicals can degrade eye health and vision slowly over the time. Exposure to increased noise, flood light sources, global warming, intense infra-red and UV radiations too damages our vision.*" (reviewed in Gupta and Muthukumar, 2018).

One lifestyle factor that is obvious is the ubiquitous use of cell phones. Prolonged fixation at a short viewing distance (see accommodation in the first chapter) has been associated with a higher risk of developing myopia (McCrann et al., 2020). In fact, any near vision activities with long periods of fixation such as reading, using computers, playing video games, all can contribute to the development of myopia (Coviltir et al., 2019; Huang et al., 2015). While studies are at times contradicting there is a suggestion that exposure to natural light during early development has been associated with reduced myopia (Sherwin et al., 2012). The proposed mechanism is that light-induced release of nitric oxide and dopamine prevents the elongation of the eye along its horizontal axis, the main cause of myopia (McCarthy et al., 2007). Addiction to screens such as smartphones and computers also means exposure to artificial light in a manner that can potentially disturb the day-night (circadian) cycle, and as such impair sleep quality. We mentioned just above with smell and hearing that poor sleep can lead to chronic stress and related diseases. Lastly, when we fixate screens with undeterred attention, we tend to blink less, which can cause a painful irritation and even inflammation of the cornea known as dry eye disease.

Another lifestyle factor that affects our vision is the change in our eating habits, mainly the easy access to relatively cheap processed food. In the 1970s, the US government subsidized corn production, which by products are used in almost all processed food. Here is a challenge: try to find any processed foods that do not have corn-derived ingredients. A major one is oil. The problem arises when considering its ratio of omega-6 to omega-3 fats. A healthy ratio is estimated to be around 5 to 1. However, corn oil has a ratio of 47 to 1. Eating fresh corn is of no or low concern compared to eating processed food spiked with the purified corn oil. The other ingredient purified from corn is the fructose (a type of sugar); we shall talk about it in the section below on taste. The incentive to mass-produced corn for transformation is money. To be enlightened on the social consequences, I recommend the documentary "King Corn" (2007). Two friends, Ian Cheney and Curt Ellis, purchased one acre of subsidized corn crop to have a glimpse at what drives the American fast-food nation. They followed their crop to where chemical magic happens. So why is an abnormally elevated omega-6 to omega-3 ratio a concern for vision? In the first chapter, we showed that the initial step of vision, phototransduction, involves membrane-bound events. One specific omega-3, DHA, comprises half of all the fat in these membranes, contributing to the fluidity of the photoreceptor outer segment disc membranes, which ensures rapidity of the process. DHA is evolution's solution to a natural anti-freeze: cold water fish such as salmon and mackerel depend on DHA for their survival. When it comes to vision, time is of the essence. Curiously, evolution of the retina has led to an awkward situation where the most oxidizable fatty acid (DHA) bathes in one of the highest concentration of oxygen of the human body. Do you see the problem here? To mitigate this situation, the retina recruits two powerful anti-oxidants: lutein and zeaxanthin. They render an orange like colour to the retinal area responsible for high acuity, the macula. Sadly, processed food does not contain these two crucial antioxidants. We get them by eating coloured whole food such as berries. Our ancestors ate fish and foraged for berries; therefore it is likely that they had a healthy vision. Of note, as we age, there is a decrease in the production of the proteins that bind lutein and zeaxanthin; therefore, even though you take supplements, only a limited amount will bind to the macula.

We will get back to vision in the section below devoted to smartphones and social media.

Tasting:

We have evolved to strive for salty foods, sugary fruits and animal fat, all of which were only sparsely available and required spending energy and organizing socially to access them. According to professor of nutritional sciences Paul Breslin, the type of food we crave can be traced from our common ape ancestors dating millions of years ago; a time when various ape species co-existed in equatorial forests, sharing fruits available all year long. When encountering a tree harbouring ripened sweet fruits, apes would climb up and binge until depletion, forcing the plump epicureans to carefully make their way back on earth, where much less caloric sour insects and leaves awaited. Hence, a healthy balance was ensured. Psychologist Deirdre Barrett proposed that purposely-enhanced artificial stimuli govern our behaviour as powerfully as that of other animals. In our so-called developed world, the stimulation of our sense of taste by ubiquitous access to junk food represents an exaggerated stimulus to cravings for salt, sugar, and fats (Barrett, 2010).

We mentioned above that one of the mass-produced corn by-products is fructose. The high-fructose corn syrups market in the U.S. is estimated at US$1.3 Billion in the year 2020 and could reach 5 billions in 2027 (https://www.reportlinker.com/p05956961/Global-High-fructose-Corn-Syrups-Industry.html). Fructose costs more than sucrose (regular sugar) but tastes much sweeter, so using it as a sweeter turns out to be more economical. The problem is that we enjoy the taste and cannot tell the difference with sucrose but we still feel hungry, why? The reason is that our body cannot use fructose to produce directly available energy, but rather stores it as fat.... so we eat or drink more to stop being hungry, and we get fatter...

How about artificial sweeteners? They taste sweet but contain no energy; therefore you should not get fat, right? We need to bring in another sense: interoception. Sensory receptors are distributed throughout our body with the role to maintain homeostasis, that is, equilibrium and integrity of our organs. It turns out that receptors for sweet taste are located not only on our tongue but also in our guts. One of their roles is to modulate hormonal signaling to the brain. Recent findings show that when infants eat artificial sweeteners, their taste preference for sweet food is increased (reviewed in Yunker et al., 2020). Brain imaging results conclude that ingestion of artificial sweeteners activates different regions than sugar, namely the ones related with rewards. The outcome is that we get rewarded for eating more. Through a literature review and statistical analysis of the numerous results on the effect of artificial sweeteners compared to sugars, Jarrett Walbolt and colleagues (2020) concluded that their consumption is associated with pre-diabetes, a condition also known as metabolic syndrome, typically characterized by obesity. Remember the 100 trillion microbes living in our guts? If you transplant them from obese mice to normal mice, guess what happens? Even when eating a healthy diet, the host mice become obese. While the exact mechanisms still elude us, there is a link with the fact that intake of artificial sweeteners leads to less numerous and diverse microbiome.

Touching:

One of the most impactful environmental changes over the past 20 years is the prominence of smartphones allowing texting. Using EEG recordings, Anne-Dominique Gindrat and colleagues (2015) observed that "*repetitive movements on the touchscreen reshaped sensory processing from the hand and that the thumb representation was updated daily depending on its use*" They proposed "*that cortical sensory processing in the contemporary brain is continuously shaped by the use of personal digital technology*".

As we rely more and more on the Internet for social interactions, we tend not to touch each other as much. We live in an increasingly virtual world, in which most of what required in person contacts now can be achieved with one click on the computer. Evgeny Morozov in his book "To Save Everything, Click Here" encourages us to slow down so we can be mindful of the deleterious implications of our ubiquitous convenience. When is the last time you wrote someone a letter? How about going to the bank? Large food chains encourage shopping using the Internet. You can order almost anything without having to leave your home. There is an apparent paradox, while loosing touch with people in the flesh, we also have become so dependent on each other's expertise; we must have faith in others and in the blind societal monster we created. According to design guru Donald Norman (quoted in Morozov's book): *"The ideal system so buries the technology that the user is not even aware of its presence."* Morozov comments on the quote as follow: "*But that means that we are more and more dependent on designers to decide what is important — or not. And they tend to emphasize the value of convenience*". He ends: "*Our friendly user interfaces seduce us to be manipulated or allow us to be just plain careless. The design gurus are responding to the market, and they will continue to nudge us in the direction of convenience*". I am not the only one saying it, and I will repeat, in my opinion the two greatest threats in our society are ignorance and convenience. Is there anything more convenient than a smartphone?

"*A smartphone is an addictive device which traps a soul into a lifeless planet full of lives*" Munia Khan

Functional brain imaging of people diagnosed as having "smartphone addiction" revealed differences in the medial prefrontal, occipital, temporal, and anterior cingulate cortices, in temporoparietal regions, and in the cerebellum (Schmitgen et al., 2020). These brain activity signatures are characteristic of addiction in general. Hence, these results support that the compulsive use of smartphone is very much a form of physiological addiction. As is the case with addictions in general, there is a close link with depression (Huckins et al., 2019).

Social media:

In the recent documentary-drama "The Social Dilemma", directed by Jeff Orlowski (2020), we learn that the product being sold by social media is not simply us, but rather the gradual, slight, imperceptible change in our own behaviour and perception; this is the resource being extracted. The change in our behaviour is gradual, and it is worth serious money considering the number of users: in the order of billions. Companies sell certainty. To achieve this, they need humongous amounts of data. The social media market trades in human future. The most valuable resource is data, which allows prediction. What kind of data? Which images you look at, how long you look at them, when, all these data are fed into algorithms that predict your future behaviour. Is this reminiscent of a protagonist from the fourth chapter? A hint, he was one of the first to put forward that our brain builds mental models from sensory inputs in order to predict the future; chillingly analogous to these new algorithms if you think about it. What would Kenneth Craik think about these models? Perhaps the same as the young geniuses who built them, then realized how they were being used, and more worryingly, that nobody has any clues on how these algorithms have learned and evolved. Cathy O'Neil warns us in her book "Weapons of math destruction: How Big Data Increases Inequality and Threatens Democracy" that major decisions impacting our lives — type of education, type of loans, cost of health insurance — are actually made by unregulated and out of control mathematical models. These black boxes bolster discrimination. For instance, mathematics rely in part on a student's zip code to determine eligibility for a loan, and consequently impacts on the type of education she-he-they will get; the poor stay poor and the rich get richer. As put forward by Michael Kearns and Aaron Roth in their book "The Ethical Algorithm: The Science of Socially Aware Algorithm Design" one of the most pressing need is to tame the beast; that is, the mathematical models that violate the basic rights of individual citizens. They argue that we must go beyond tackling laws, regulations and establishing watchdog groups: we must replace the algorithms themselves with "socially aware algorithm designs".

As it currently stands, we are pointing artificial intelligence tools back at ourselves to see what happens, and this is out of control. Companies use artificial intelligence to personalize interfaces for each user: how to give them a dopamine hit. Social media is a drug, it banks on dopamine bursts, and creating dependency... it relies on our innate needs to socialize: dopamine deficit states are painful. We are hooked. The Internet is manipulating you, it wants things from you, it is making you addicted, and social media is using your psychology against you. In a way, Internet companies are like drug dealers. In his book, "Superintelligence: Paths, Dangers, Strategies", Nick Bostrom warns us that artificial intelligence is becoming out of control and surpasses human intelligence. However, through this tedious read (not an easily accessible blend of philosophical and technical jargon), one does get a feeling of hope, if we act now — such as by implementing "socially aware algorithm design" — we could engineer initial conditions that minimize the deleterious impact of the artificial intelligence explosion, and optimize its adaptative integration to our evolution as a society. In theory, we could reverse engineer behavioral addiction to harness these dopamine boosters to help us better communicate with each other, find a healthier balance between work and play, and overall guarantee health and happiness for our children (recommended read: "Irresistible: The Rise of Addictive Technology and the Business of Keeping Us Hooked" by Adam Alter).

Data from our senses and related behaviours, that is, how we act on social media, are used to make a more and more accurate predictor model. Here is an irony: some predictor models are based on artificial neural networks. Companies can then personalize advertisements dictated by the predicted behaviours, even more: there is a third party who is paid to manipulate individuals, each individual (yes each Internet user). Each individual being manipulated has his own price. We can demand not to be an extractable resource!

Facebook is in charge of your newsfeed the same way a magician creates illusions. Everybody gets different news, information, which exacerbates political polarization. Because false news spread 6 times faster, the algorithms rely more on them to increase profit. We have created a monster. As Nick Bostrom predicted in his 2014, out of control artificial intelligence, with its stealth algorithms, exerts an ungraspable ubiquitous control on social media; ungraspable in the sense that nobody knows how these algorithms have evolved, yes evolved, because they learn and change. One of the outcomes is that people believe more and more in conspiracy theories and act accordingly, concretely... this is the cost to society. Furthermore, there is a tendency for people to read only titbits, to be rapidly distracted, without taking the time to read long articles and let alone checking the sources (recommended read: "The Shallows: What the Internet is doing to our brains" by Nicholas Carr, 2011).

Quoting Bruce Hood from his book "The Self Illusion", which we discussed in the fourth chapter: "*With their combinations of distorted opinions, rapid communication without time for reflection, and the perceived distance and anonymity, social networks are the perfect platform for us to behave in a way that we would not in real life*".

A large-scale study has established a link between problematic social networking site use and psychiatric disorder symptoms, particularly in adolescents (Hussain and Griffiths, 2018). Numerous negative factors are involved with the compulsory use of social networking sites. There is a lack of face-to-face communication, a lack of creativity, and sadly a lack of self-esteem (Bergagna and Tartaglia, 2018) alongside an unfavourable body image perception, particularly in females, due to frequent body and weight comparisons and the use of software to alter photos (Eckler et al., 2017). Not surprisingly, adolescents addicted to Facebook showed increased rates of social anxiety, generalized anxiety, and depression (Muzaffar et al., 2018). One can list many other negative factors: 1) waste of time; 2) abandonment of family; 3) privacy disrupted; 4) insomnia; 5) physical inactivity; 6) internet addiction; 7) cheating; and 8) moral corruption. The biggest issue with social media is cyber bullying, which as we know has driven people to dying by suicide (Messias et al., 2014).

Social media becomes an extension of the self, even becomes the self. Our predictive mental models relate to the Internet world, navigating its environment, interactively, eliciting survival mechanisms and focusing on reproduction with actual physiological reactions, from a brain wired to survive and reproduce. However, because the environment is not real, it does not pause an actual threat and does not directly allow reproduction itself, but nevertheless it very much elicits both reactions, viscerally.

To borrow Munia Khan's metaphor, the Internet is a "lifeless planet full of lives". In his book about this lifeless planet full of lives, "Picnic Comma Lightning (the experience of reality in the 21st century)", Laurence Scott paints a grotesque spectacle of having access to an infinite number of other people's realities through the lens of decaying mental models of the world. Our own realities intricately percolate in this lifeless planet ruled by mysterious algorithms that undermine our attempts at self-definition through their own data-driven portraits where private and public have no boundaries. One of the founders of virtual reality himself, Jaron Zepel Lanier, strongly encourages us to delete all of our social media accounts, right now (Lanier, 2018).

Let us end on a positive note. The Internet has its uses, quite so, to the point of changing how we might evolve as a society. If you have not read Marc Prensky's book, "Brain Gain (technology and the quest for digital wisdom)", he describes how in principle digital technology can complement what our brain cannot achieve on its own. The wise use of digital technology, which he calls "digital wisdom" has the potential of improving how we learn, communicate, invent, and even evolve. Prensky stresses the need to educate kids early, using what he calls "imag-u-cation". The term originates from the crucial role to educate each student to use her-his-their imagination as a mean to learn digital wisdom. Very briefly, the teacher would ask questions such "*If each of us had an iPad (or a laptop, or an iPhone, etc.) at this moment, or if our class had access to a supercomputer, or any other kind of technology, what could we do with it that would enhance our learning in a major way? What would be the digitally wise way to use these technologies? What use would be merely clever, or dumb?*". I would add that if we want to indeed evolve as a society, we need introspection, serious introspection and yes education, serious education related to moral values, humane behaviour and finally this word again: "compassion", compassion for others and for ourselves as individual. One example of self-compassion is to set time aside to breathe, or pray, or meditate, or contemplate; call it what you want but set time aside for you. Personally, I connect to my primary senses by observing my breathing. Closing your eyes in silence and stillness is the best way to truly take in the reality your brain has constructed.

Concluding Remarks

"*The world's continual breathing is what we hear and call silence.*" Clarice Lispector (The passion according to G.H.)

Congratulations, you have a brain, a brain that can do so much: generate a reality from your senses or without. We are just barely starting to understand how it does all that, how it evolved from single cells 3.8 billion years ago. You possess one; infinity is in you. Look no further; count your riches. Breathe in the infinite beauty of you, be kind, be curious, and be now.

How about all the suffering in the world? What is the point of reading this book if you don't even have an answer to the question: what is the meaning of life? Who will read this book anyway? OK, you are reading it, or part of it. There are so very many books out there, famous communicators, so many organizations, groups, etc. This book is a seed, one of many, planted carefully by an essentially unknown person, intentionally, to germinate in a few, perhaps you? Perhaps today you will smile at someone, a smile of calmness, a gentle smile. Perhaps you will start a conversation with someone, realizing they believe in some form of pseudoscience. I would recommend listening to them, and in an unthreatening way, talk about what you might have in common. Hey, how about the fact that we are blind about 4.5 hours every waking day (see first chapter)? How about trying to explain that, just that. Maybe also bringing the fact that when you look at yourself in the mirror, you cannot see your eyes moving. Talking about evolution should not, in theory, be threatening, but somehow it is just a belief for some, which they chose not to believe. Maybe you can talk about critical thinking (see the seventh chapter) in the most humane way, again non-threatening, and most importantly non-judgmental manner. What you might have in common with many is feeling some level of anxiety, at times sadness, loneliness, etc. This is a common point. Maybe, again many maybes, you could talk about breathing, the feeling of breathing in silence.

Breathe in silence

We need silence and solitude, (taking time to do absolutely nothing, to need absolutely nothing, even only 5 minutes per day), some call it prayer, relaxation, meditation, or doing nothing, during this time you are letting go at this moment, letting go, you observe your breath, (meditating is a synonym of observing), you let go to your breath, you pay attention to the place in your body where you can feel it (belly going up and down, movement of air through your nostrils, other places where a sensation follows the rhythm of your breath), the rhythm of your breath, letting go, letting go, to a precious moment when breathing has a mind of its own, when breathing just happens, trusting that breathing does not require any action from your conscious will, letting go, letting go.

Breathing is an automatic action of the body controlled by a region of the brainstem (PreBötzinger Complex) that coordinates all the nerves and muscles creating this rhythm and the final outcome essential to life: breathing (discussed in the sixth chapter). We can breathe without the need to think about it.... and when we do and accept that there is nothing required from us, and we let go to this automatic rhythm and function, then we can be in a state of peace; in a way, one can say they found bliss, solace, enlightenment, perhaps even God. If someone reaches such states by observing their breathing, then the source is the PreBötzinger Complex, part of the "reptilian brain". The generation of the feeling of bliss occurs in the limbic system (decreased threat related to the amygdala and improved memory related to the hippocampus) as well as the cortex (posterior cingulate cortex; prefrontal/frontal cortex; and insular cortex). In fact, most regions are interconnected; therefore the feeling of bliss reflects inter-connectedness rather than independent activation of specific brain regions.

Am I desacralizing, profanating? No, I am amazed, I contemplate the beauty of how we are, I am curious, I feel no threat... I let go, I do not need to do anything at this moment, I am energized. As I start to question, my breathing changes, and changes again. It makes me question more, I start to feel a threat (we are threat detectors, and we rely on interneurons to create mental models), a perceived threat, now my breathing requires will, effort, this is my impression, I no longer let go... my thoughts are racing. Then I accept that thoughts are also a product of my brain, and I observe them as I observe my breathing, and let go, realizing that there is no threat, that I can simply be here and now... alone.... 5 minutes.... precious 5 minutes... and I repeat and repeat every day 5 minutes... I let go... I found the sacred in my life. I am infinitely blessed.

Breathing with the leap of faith

Waterboarding is torture, it is a mock execution; victims will suffer unfathomable traumatic scars for the rest of their lives (Xenakis, 2012). All is set to make you believe you will drawn, your senses, the context of torture, together tell you: you cannot breathe. In fact you can, in theory, but won't commit to breathing: you are being executed. There are moments when we think it is over, no way out; death perhaps is the only outcome. We do not realize that we can actually breathe; we need a leap of faith. Like waterboarding, our senses tell us we cannot breathe. However, with this one leap of faith, air comes in, the feeling of air is felt, we breathe, we are content with the breathing, only breathing, how so very counterintuitive. The current context is overwhelming: we should not be able to breathe. But we are. Can we simply accept it? Pretty hard, extremely hard, impossible, our body is on alert mode and all our senses say we cannot breath; except, the feeling of air coming in if truly felt as it can, says otherwise. We focus on breathing, we realize against all odds, we are breathing, we are alive, and you know what, we do not need anything else, we start looking around to assess threats (as we are wired to do) and take time to absorb what are real or perceived threats. Be curious about what you observe.

Be curious

We mentioned that being able to ask the question "what is the meaning of life?" is an offshoot of the evolution of predictive mental models, and that the urge to ask could be related to the fact that we have no sensory inputs from the cells that generate such models. Since the question itself is consequential to 3.8 billion years of evolution, the answer becomes irrelevant. However, this should not stop us from asking this question and many others.

In their peer-reviewed paper entitled "Did curiosity kill the cat? Evidence from subjective well-being in adolescents", Veljko Jovanovic and Dragana Brdaric (2012) link curiosity with giving a meaning to life. They reported that "... *adolescents high in trait curiosity have higher levels of life satisfaction and positive affect and greater sense of purpose in life and hope than adolescents with both low and average curiosity*".

There is so much we can learn, so much we do not understand yet. For instance, we are just starting to decipher the mechanisms underlying how our brain generates the various predictive mental models, some remaining in our unconscious, others reaching consciousness. Other examples include social inequities and voluntary harm to others. We need to learn about history in order not to repeat it. What can we do to prevent this, to stop this? What can we do to prevent the disastrous consequences of human-caused climate changes? How come only 2% of our genes are used to encode proteins (exons), what about the other 98% (introns), which were until very recently dismissed as junk? We are just starting to understand the role of introns in controlling gene expression, their involvement in genetic diseases, viral infections, sex determination, and evolution (Carey, 2015). As humans, we have a unique capacity for abstraction. We can think about concepts that have nothing to do with our senses. It can be a curse (endless rumination, worries, false threats) or a blessing (constructive curiosity). In chapter one we mentioned the first "photo" of a black hole (M87), stressing that it was a mathematical construct. We cannot use any of our senses to study black holes: any form of energy looses its definition inside black holes. So we rely on technological advances to transduce the forms of energy that surround the black hole, just prior to their vanishing.

There is overwhelming evidence linking curiosity with feelings of happiness and self-content (Hausler et al., 2017; Lydon-Staley et al., 2020). Curiosity can also help you overcome anxiety; you are "*more likely to engage in approach behaviour amidst conflicting avoidance motivation*" (Kashdan et al., 2008). Cultivating curiosity has been associated also with increased lifespan (Swan and Carmelli, 1996).

Talking about being curious. Was there an evolutionary advantage to curiosity? From a purely selfish perspective, if there was an advantage, what did it provide or may still be providing? It would make me very happy, less anxious and might extend my longevity if we could keep evolving as a society. There is converging evidence suggesting that curiosity related behaviours are evolutionarily ancient tendencies that involve the activation of dopaminergic motivational and reward systems. Curiosity has played a crucial role in the survival of species. The urge to explore novel aspects of the external and internal environments assists in maintaining vigilance and in acquiring knowledge about the changing environments. It might not come as a surprise that brains evolved to release dopamine and other feel-good chemicals, such as oxytocin, when encountering novelties (Di Domenico and Ryan, 2017). As we learn more about the link between curiosity and the activation of our reward system, we develop innovative approach to teach students. The more you are curious, the more you learn and the more you learn, the more questions arise, hence, the more curious you become. We can tap into this positive-feedback loop. Ultimately, we can come up with intelligent computational models designed to foster curiosity and learning (Oudeyer et al., 2016). This is good news; does it make you want to smile?

Fuelled by curiosity and the associated hormonal reward, we learn by establishing novel mental models; the very models put forward by the pioneering work of Kenneth Craik, as we discussed in the fifth chapter. Neural representations of the external and internal environments are constructed and updated in a manner that depends on behavioural context, that is, it can be fostered by curiosity. Erin Glennon and colleagues (2019) showed that the neurotransmitter noradrenaline (produced in the locus coeruleus) improved motivation and perseverance to be curious as assessed with exploratory behaviour in rats. An important sensory modality for rodents to explore their environment is by moving their whiskers. Results from Goren Gordon and Ehud Ahissar (2012) support that curiosity starts by rodents constructing a mental model of whisker motion, which then extends to models of how the air movement and the touching of objects alter the mental models of whisker movements.

What are the neural networks that are responsible for being curious? We know that noradrenaline contributes to being curious and that dopamine and oxytocin both reward curiosity. Most research is being done by quantifying exploratory behaviour in animals such as rats and mice. Anthony T Lee and colleagues (2019) found that interneurons containing the molecule VIP (vasoactive intestinal polypeptide) played an essential role in mice trained to perform a standardized behavioural test called the "elevated plus maze". In brief, picture an elevated platform in the shape of a cross where two facing arms are enclosed by walls (closed arms) and the two others are open (open arms), which means more danger to fall. Anxious mice tend to spend more time in closed arms, while mice motivated to explore (to be curious) spend more time in the open arms. Lee and colleagues found that VIP interneurons were more active when mice stood on the open arms. The circuit itself consists of VIP interneurons releasing the inhibition of the prefrontal responses to hippocampal inputs that are known to send signals to avoid open arms. In other words, VIP interneurons alter the predictive mental models that guide complex behaviours, in the case of this behavioural test, enticing exploration, or from a human extrapolation, being more curious, less anxious, and perhaps "more happy". To be clear, I am extrapolating, and not saying that being happy or not applies to mice, this would be a dangerous slope toward anthropomorphism. Let us restrict our conclusion to the fact that VIP interneurons participate in creating mental models that contribute to making mice more curious.

Ultimately, being curious is about taking care of you. In addition to being associated with happiness in humans, curiosity improves memory (Gruber and Ranganath, 2019), motivates you to learn (von Stumm et al., 2011), helps strengthen relationships (Todd et al., 2004) and boosts your self-esteem (Macaskill and

Denovan, 2014). Curiosity can motivate you to find meaning in experiences, including those related to yourself.

Be kind

Throughout the book, I used the word "compassion" but never "empathy" until just now. According to Paul Bloom (2017), empathy "*is narrow in its focus, rendering it innumerate and subject to bias. It can motivate cruelty and aggression and lead to burnout and exhaustion. Compassion is distinct from empathy in its neural instantiation and its behavioral consequences and is a better prod to moral action, particularly in the modern world we live in*". Before the book ends, it might be a good time to define the word "compassion". Clara Strauss and colleagues (2016) propose five elements to characterize what being compassionate entails: 1) recognizing suffering; 2) understanding the universality of human suffering; 3) feeling for the person suffering; 4) tolerating uncomfortable feelings; and 5) motivation to act/acting to alleviate suffering. It is on the basis of such operational definition that I deliberately chose to use the term compassion throughout the book. Let us reflect on its evolutionary context. Jennifer L Goetz and colleagues (2010) put forward that compassion evolved from the need "*to facilitate cooperation and protection of the weak and those who suffer*". In essence, compassion is a form of social glue and its manifestations differ between cultures, while respecting the five elements mentioned above. The neural basis remains a matter of debate. A tempting explanation is based on the accidental discovery of so-called "mirror neurons" by the Italian neuroscientist Giacomo Rizzolatti (Rizzolatti et al., 1996). While recording the activity of cells involved in planning and executing bodily movements, they observed firing from these neurons when monkeys were looking at a researcher reaching for raisins. Motor neurons from primates fired while they observed a human doing an action. In 2010, the neurosurgeon Roy Mukamel and his team managed to record similar occurrences in subject operated to treat severe epileptic seizures (Mukamel et al., 2010). However, evidence is lacking to assert that humans have similar mirror neurons, let alone that they are involved in compassion.

Here is a fascinatingly pertinent evolutionary aspect proposed by Mark Changizi and his colleagues in 2006. It has to do with red cones and exposed skin. Let me explain. Among mammals, only humans and some primates (old world monkeys, including baboons and macaques) are trichromates, that is, their retina contains blue, green and red light sensitive cones. What they tend to have in common is the lack of facial hair. Even humans with beards still have some facial skin exposed. Changizi put forward that the ability to see the red component of skin colours provides information about both the emotional and health statuses of an individual. More recently, Chihiro Hiramatsu and colleagues (2017) shared experimental evidence that indeed the addition of red cones (to the blue and green cones) optimally assists old world monkeys in detecting social facial colour signals. You might ask about humans with high levels of melanin (darker skin). Interestingly, it depends on whom you grew up with. Individuals with fair complexion, raised by individuals who also have a fair complexion have more difficulty recognizing emotions and health status of darker complexion individuals. However, if raised by dark complexion individuals, then, those with fair complexion can easily recognize emotions and health statuses in individuals with darker skin (Changizi, 2009).

How about compassion for yourself, being kind to you? We talked about mindfulness, where you pay attention to your breathing; in fact I spent quite a lot of time referring to connecting to one's sensation of breath. An offshoot of mindfulness-based meditation is compassion meditation. Here you apply compassion to you. There is indication of a positive effect on the neuroendocrine system and immune responses (Pace et al., 2009). Brain imaging in people who practice compassion meditation shows the functional modulation of regions involved in compassion associated with true altruistic caring responses; such changes entail increased neural activity in the inferior frontal gyrus and the dorsomedial prefrontal cortex (Mascaro et al., 2013). Another research team studying expert practitioners of compassion-meditation found increased activation in brain regions involved in affiliation, positive affect and reward processing including the ventral striatum and the medial orbitfrontal cortex (Engen and Singer, 2015). In addition, training people to have deliberate compassion for others was associated with changes in the activity of distinct neural networks: the medial orbitofrontal cortex, putamen, pallidum, and ventral tegmental area; these are all brain regions previously associated with positive affect and affiliation (Klimecki et al., 2013).

Be compassionate to others and yourself, be now.

Be now

"*I' is merely one of the world's instantaneous spasms.*" Clarice Lispector (The Passion According to G.H.)

"*As soon as you discover the truth it's already gone: the moment passed. I ask: what is it? Reply: it's not.*" Clarice Lispector (The Hour of the Star)

"*I am obscure to myself. I let myself happen. I unfold only in the now. I am rudely alive.*" Clarice Lispector (Água Viva).

"*I almost live, therefore, in the past because I can't recognize the type of richness of the present moment. The forgetting of things is my escape valve. I forget a lot out of necessity. I'm even trying and succeeding in forgetting me, me minutes before, I forget my future. I'm naked.*" Clarice Lispector (A breath of life)

"*You must live in the present, launch yourself on every wave, find your eternity in each moment. Fools stand on their island of opportunities and look toward another land. There is no other land; there is no other life but this.*" Henry David Thoreau (Walden)

"*Has it ever struck you that life is all memory, except for the one present moment that goes by you so quick you hardly catch it going?*" Tennessee Williams (The Milk Train Doesn't Stop Here Anymore)

This is where and when everything begins. Let go.

I smile because I am happy because I smile because I am happy I smile because I am happy because I smile because I am happy because I smile because I am happy because I smile because I am happy because

Acknowledgments

I would like to thank the many friends who ventured into reading various parts and stages of this book, and provided pertinent comments. These are: Etienne Fillion-Sauvé, Michael Tremblay, Andrée Caron, Marc-André Roy; Manuel Vidal-Sanz; Jim Harley Marsh; Sharee Kuny, Ken Cockwill; Rick Schulz; Yves Parent, and Frédéric Gaillard. I thank the late Dr. Thomas Alfredo Reader for contaminating me with his passion for neuroscience and Dr. Michael Rasminsky for transmitting his scientific rigour and commitment to writing with utmost effectiveness and clarity.

Finally I want to make it very clear that my inspiration and gratitude remains toward the student I guided and taught. I think of them often; they are the reason I remain curious and passionate about life in general. A few names come to mind: Doha, Jill, Camille, Ted, Mathew H, Matthew B, Greg, Paul, Antonia, Ioannis, Mandy, Rachel, Emma, Marc-André, Blake, Jonathan, Isha, Tanner, Allan, Heather, Lukas, Aaron, Adam, Cale, Dylan, Eric, Ifrah, Muhamad, Linzi, Jiyeon, John C, Justin, Lakhan, Katherine, Kristen, Melania, Neel, Nicholas, Preet, Qian, Rami, Samiha, Shveta, Stephanie, Shubam, Tiffany, Robert, Ronja, Valerie, Victor, Yang, Jashanjit, Jarlo, Alexander, Ali, Christian, Dale, Danae, Danielle, Doyeon, Eyram, Jennifer, Kassandre, Margaret, Mizuki, Davey, Bijal, Vladimir, Silvina, Bradi, Brittany, Beatrice, Winnie, Carina, Carlos, Claire, Clark, Devan, Erin, Geoff, Gordon, Hope, Jason, Julia, Jyoti, Luis, Maryam, Monique, Paige, Parnian, Sara, Shuo, Stacey, and Caleb.

Figures (Legend and Source)

Book cover. "The meaning of life", written in various languages; drawn by the author based on translations using "Google translate".

Figure 1. Circular reasoning. Based on illusions created by Akiyoshi Kitaoka, who built these based on the Popple phase-shift illusion (Popple and Sagi, 2000), itself based on the Fraser illusion (Fraser, 1908). Image name: Asymmetrical Circles. Source: https://gpuzzles.com/optical-illusion/asymmetrical-circles/. Contact info: info@gpuzzles.com. Composition drawn by the author.

Figure 2. Café wall Illusion, also known as the Munsterberg illusion or "The Shifted Checker-Board-figure" (Pierce, 1898). Author: Fibonaci; 15 March 2007, 08:44 (UTC). Description of the Café wall illusion: the horizontal lines are parallel, even if they seem otherwise. Permission is granted to copy, distribute and/or modify this document under the terms of the GNU Free Documentation License, Version 1.2 or any later version published by the Free Software Foundation; with no Invariant Sections, no Front-Cover Texts, and no Back-Cover Texts. Image source: https://upload.wikimedia.org/wikipedia/commons/d/d2/Café_wall.svg. Name: Composition drawn by the author.

Figure 3. Magnet Illusion. Also called the blivet illusion, impossible fork, trident illusion, or Schuster's fork. This illusion combines two incompatible 3D perspectives at once, creating a "lost" layer between the top two rods, and an impossible extra, vanishing rod in between the bottom two. Its origin can be traced to a 1964 article by D.H. Schuster in the American Journal of Psychology: Schuster, D. (1964). A New Ambiguous Figure: A Three-Stick Clevis. The American Journal of Psychology, 77(4), 673-673. Drawn by the author.

Figure 4. Visual illusions are real. Drawn by the author.

Figure 5. Light pathway to the retina. Drawn by the author.

Figure 6. Photon reaching a foveal cone. Drawn by the author.

Figure 7. Photon reaching a cone outer segment disc. Drawn by the author.

Figure 8. Photon absorbed by the chromophore, retinaldehyde. Drawn by the author.

Figure 9. Before photoisomerization. Drawn by the author.

Figure 10. After photoisomerization. Drawn by the author.

Figure 11. End of light-dependent events in vision. Drawn by the author.

Figure 12. A matter of perception. Drawn by the author.

Figure 13. Phototransduction. R: rhodopsin; R*: activated rhodopsin; T: G-protein transducin; GDP: guanosine diphosphate; GTP: guanosine triphosphate; $T\alpha$*: activated transducin; PDE: phosphodiesterase; PDE*: activated phosphodiesterase; cGMP: cyclic guanosine monophosphate; GMP: guanosine monophosphate; CNG: cGMP-gated nucleotide gated channel; NCKX: sodium, calcium, potassium ion exchanger. Drawn by the author.

Figure 14. Dark current. Drawn by the author.

Figure 15. Outcome of dark-current shutdown on photoreceptors and their synaptic communication with bipolar cells. V: voltage; Ca++: calcium; BC: bipolar cell; ON BC: depolarizing bipolar cell; OFF BC: hyperpolarizing bipolar cell; mGluR6: metabotropic glutamate receptor type 6; iGluR: ionotropic glutamate receptor; H: horizontal cell; CaV: voltage gated sodium channel. Drawn by the author.

Figure 16. Retinal ganglion cells provide the only retina output. Drawn by the author.

Figure 17. Visual pathways. LGN: Lateral geniculate nucleus, V1, V2, V3, V4: Primary (first), second, third, fourth visual cortices, respectively; TEO: Posterior inferior temporal cortex; TE: Anterior inferior temporal cortex; V5/MT: Middle temporal area (fifth visual cortex); MST: Medial superior temporal area; LIP: Lateral intraparietal area; VIP: Ventral intraparietal area; STP: Superior temporal polysensory area; 7a: Visual area 7a in the parietal cortex (Brodmann's terminology). Based in part on information reviewed in Rokszin et al., 2010. Drawn by the author.

Figure 18. Hering illusion. Dates back from in 1861, attributed to German physiologist Ewald Hering. According to Mark Changizi, "*Evolution has seen to it that geometric drawings like this elicit in us premonitions of the near future. The converging lines toward a vanishing point (the spokes) are cues that trick our brains into thinking we are moving forward as we would in the real world, where the door frame (a pair of vertical lines) seems to bow out as we move through it and we try to perceive what that world will look like in the next instant.*" Drawn by the author.

Figure 19. Rotating Snakes peripheral drift illusion, Monochromatic version of illustration by CMG Lee (PNG file at 2000px) based on a design by Kitaoka Akiyoshi. Source link: https://commons.wikimedia.org/wiki/File:Rotating_snakes_peripheral_drift_illusion.svg

Figure 20. A short history of the senses. Drawn by the author.

Figure 21. Phototaxis and chemotaxis in unicellular organisms. SRs: sensory rhodopsins (photoreceptor); Htrs: cognate transducer proteins; MCP: methyl-accepting chemotaxis protein (chemoreceptor); MLP: MCP-like protein; P: phosphoryl group; Pi: inorganic free phosphate; CH3: methyl group; ADP: adenosine triphosphate; ATP: adenosine triphosphate; CheA: histidine kinase; CheB: methylesterase; CheR: methyltransferase; CheW: adaptor protein; CheY: autokinase; CheZ: phosphatase; CW: clockwise. Author's redrawing of figure 1 from Suzuki et al., 2010.

Figure 22. Ocelloid from *Nematodinium*. Drawn by the author after Kofoid and Swezy (1921) and Francis, 1967.

Figure 23. Ctenophorae: Public domain due to publication date exceeding 100 years (1904). Ernst Haeckel (1834 - 1919) - Kunstformen der Natur (1904), plate 27: Ctenophorae. Legend: Haeckelia rubra (Victor Carus, 1862) = Haeckelia rubra (Gegenbaur, Kölliker & Müller, 1853), from above Haeckelia rubra (Victor Carus, 1862) = Haeckelia rubra (Gegenbaur, Kölliker & Müller, 1853), from the side Hormiphora foliosa (Haeckel) = Hormiphora sp., from the side Callianira bialata (Delle Chiaje) = Callianira bialata Delle Chiaje, 1841, from the side Tinerfe cyanea (Chun) = Tinerfe cyanea (Chun, 1889), from the side Lampetia pancerina (Chun) = Lampea pancerina (Chun, 1879), from the side.

Figure 24. Seven fundamental transduction pathways common to all species. TGF-β: transforming growth factor beta; SMAD: refers to homologies to Caenorhabditis elegans SMA ("small" worm phenotype) and to Drosophila MAD ("Mothers Against Decapentaplegic") family of genes; SMURF: SMAD specific E3 ubiquitin protein ligase; BMP: bone morphogenetic protein; CHRD: chordin; FST: follistatin; NOG: noggin; CAN: cerberus/DAN/gremlin family; DAN/NBL1: neuroblastoma suppressor of tumorigenicity 1 isoform 1; SARA: SMAD anchor for receptor activation; SKI/SNO: SKI proto-oncogene/SKI-related gene; WNT: wingless-type MMTV integration site; FZD: frizzled; LRP5/6: lipoprotein receptor-related protein 5/6; DSH: dishevelled; GSK-3:

glycogen synthase kinase 3; b-cat: b-cat: b-catenin; TCF/LEF: T-cell-specific transcription factor/lymphoid enhancer binding factor; WIF: WNT inhibitory factor; DKK: dickkopft; CER: cerberus; SFRP: secreted frizzled receptor protein; LRP: low density lipoprotein receptor-related protein; APC: adenomatous polyposis coli; GSK3: glycogen synthase kinase 3; NLK: nemo-like kinase; TCF/LEF: transcription factor/lymphoid enhancer binding factor; NR: nuclear receptor; FA: fatty acid;; MIB: mindbomb; NEURL: neuralized; PEN2: presenilin enchancer 2, g-secretase subunit; PSEN: presenilin-1, g-secretase subunit; APH1: anterior pharynx defective homolog 1, γ-secretase subunit; ADAM: a disintegrin and metalloproteinase domain; DTX: deltex; NLE: notchless; *O*-fut: o-fucosyltransferase; SMRT: nuclear receptor co-repressor; CSL: CBF1, suppressor of hairless, LAG-1; MAM: mastermind; HES: hairy and enhancer of split; HFN4: hepatocyte nuclear factor 4 (HNF4); RXR: retinoid X receptor; HH: hedgehog; PTC: patched; SMO: smoothened; CK1: casein kinase 1; GSK3: glycogen synthase kinase 3; COS: costal; FU: fused; GLI: glioma-associated oncogene; SuFU: suppressor of Fused; GF: growth factor; GFR: growth factor receptor; Ras/Raf: proto-oncogene serine/threonine kinase; TF: transcription factor; PTK: protein tyrosine kinase; RTK: receptor tyrosine kinase; MAPK: mitogen activated protein kinase; PLC: phospholipase C; PKC: protein kinase C; JAK/STAT: janus kinase/ signal transducer and activator of transcription; Cyto: cytokine; CytoR: cytokine receptor; EPHR: ephrin receptor; EGFR: endothelial growth factor receptor; INSR: insulin receptor; VKR: venus kinase receptor; FGF: fibroblast growth factor; FGFR: FGF receptor; TAD: transactivation domain; NFκB: nuclear factor kappa-

B; SOCS: suppressor of cytokine signaling; PIAS: protein inhibitor of activated STAT; SHPs: SH2 domain protein tyrosine phosphatase; VCL: vinculin; PXN: paxillin; ACTN: actinin; TLN: talin; ITG: integrin α/ß; FN: fibronectin; DSG: dystroglycan; colαIV: collagen αIV. Drawn by the author based on information reviewed in Babonis and Martindale, 2016.

Figure 25. An imaginary meeting. Kenneth Craik (left) and Ramon y Cajal (right). Drawn by the author.

Figure 26. Simplified neural networks underlying mental models. A) Side view of the primate basal ganglia and their associative loops. Square magnified on left for the associative loop and on right for the ventral (limbic) loop). In the left hemisphere in humans, the different shades highlight the connectivity between separate areas in the cortex and their specific targets in the striatum and thalamus. This parallel partial segregation is maintained throughout the basal ganglia in the GP (globus pallidus), STN and SNr. B) Side view of the lamprey basal ganglia and associative loops. The circuit involves BG, thalamus and pallium, which projects directly towards the striatum and in turn receives mediated (via thalamus) inhibitory output from the BG. C) Front view of the drosophila central complex and associative loops showing connections between the protocerebral bridge (PB), fan-shaped body (FB) and ellipsoid body (EB), along with two satellite neuropils, the Gall and lateral accessory lobe (LAL). The noduli have been omitted. The PB is divided into symmetrical synaptic modules. PB modules encode sensory representations (visual/tactile) from eight sectors on each side of the animal's long axis. Left and right representations of this 'where' code are relayed across eight modules of the FB such that left and right maps from the PB are compared. The grey and bold black fonts indicate modules on the left and right side, respectively. Outputs from the FB project into 16 modules of the EB such that modules representing opposite sectors are adjacent. The PB receives numerous inputs, including neurons entering laterally (arrows) carrying high-level information about visual motion direction. Columnar module in the FB are intersect by many dendritic trees and terminals (two shown) that likewise carry synthesized sensory information about complex parameters ('what' inputs) as well as a broad palette of modulatory peptides. The EB also receives a variety of inputs (also 'what' afferents), such as from the Gall and other satellite neuropils. Different combinations of where and what inputs result in different levels of activity in EB modules. Competing strengths of activity among EB modules result in a few achieving a stable output to the LAL. Further connections link the LAL system to pre-motor descending neurons (not shown). Abbreviations: PB, protocerebral bridge; FB, fan shape body; EB, ellipsoid body; LAL, lateral accessory lobes; Pal, pallium; Str, striatum; MLR, mesencephalic locomotor region; Thal, thalamus; GPe, globus pallidus external segment, GPi, globus pallidus internal segment; STN, subthalamic nucleus, SNr, substantia nigra pars reticulata; Cau, caudate; NAcc, nucleus accumbens; Put, putamen. Drawn by the author after Fiore et al., 2015; adapted from figure 1.

Figure 27. Magician, drawn by the author after, https://pxhere.com/en/photo/1448671

Figure 28. Neural networks linking the dopamine reward centre (VTA, Ventral Tegmental Area), and the regions involved in regulating mood. Note the many reciprocal connections, supporting that various regions can regulate each other, which constitutes a precarious situation amenable to instabilities underlying mental health disorders such as depression and anxiety. Abbreviations: mPFC medial Prefrontal Cortex, LH Lateral Hypothalamus, PPTg Pedunculopontine Tegmentum, LDT Laterodorsal Tegmentum, BNST Bed Nucleus of the Stria Terminalis, LC Locus Coeruleus, RMTg Rostromedial Tegmental Nucleus, LHb Lateral Habenula, Hipp Hippocampus, NAc Nucleus Accumbens, VP Ventral Pallidum, Amyg amygdala, OFC Orbitofrontal Cortex. The plus and minus signs indicate excitation and inhibition, respectively. Drawn and adapted by the author after Liu and Chaudhury, 2017; figure 14.2.

Figure 29. Autonomous control of respiration. Of note, the pre-Bötzinger complex (pre-BötC) is the hub to this closed-loop, its role is to control inspiration as part of the central pattern generator neural networks. Abbreviations: Abd. – abdominal; AbN – abdominal nerve; aug-E – augmenting expiratory neuron; BötC – Bötzinger complex; CPG – central pattern generator; cVRG – caudal ventral respiratory group; early-I – early-inspiratory neuron; late-E – late-expiratory neuron; Mns – motoneurons; NTS - nucleus of the tractus solitarius; P-cells – Pump cells; P(e) – excitatory pump cells; P(i) – inhibitory pump cells; pFRG – parafacial respiratory group; PN – phrenic nerve; post-I – post-inspiratory neuron; pre-I/I – pre- inspiratory/inspiratory neuron; PSRs – pulmonary stretch receptors; ramp-I – ramp-inspiratory neuron; RTN – retrotrapezoid nucleus;

rVRG – rostral ventral respiratory group; VRC – ventral respiratory column. Drawn by the author after Molkov et al., 2014; figure 1B.

References

Amit R, Abeles D, Carrasco M, Yuval-Greenberg S. Oculomotor inhibition reflects temporal expectations. Neuroimage. 2019;184:279-292.

Armitage JP, Hellingwerf KJ. Light-induced behavioral responses (phototaxis) in prokaryotes. Photosynth Res. 2003;76(1-3):145-55.

Aymerich-Franch L, Petit D, Ganesh G, Kheddar A. The second me: Seeing the real body during humanoid robot embodiment produces an illusion of bi-location. Conscious Cogn. 2016 Nov;46:99-109.

Azzopardi P, Cowey A. Preferential representation of the fovea in the primary visual cortex. Nature. 1993 Feb 25;361(6414):719-21.

Baaske P, Weinert FM, Duhr S, Lemke KH, Russell MJ, Braun D. Extreme accumulation of nucleotides in simulated hydrothermal pore systems. Proc Natl Acad Sci U S A. 2007 May 29;104(22):9346-51.

Badde S, Myers CF, Yuval-Greenberg S, Carrasco M. Oculomotor freezing reflects tactile temporal expectation and aids tactile perception. Nat Commun. 2020;11(1):3341.

Bagasra O, Heggen C. Hossain M. Chapter 3: Olfaction and autism. In: Autism and Environmental Factors. Hoboken, NJ:John Wiley & Sons, 83-95 2018.

Barberia I, Blanco F, Cubillas CP, Matute H. Implementation and assessment of an intervention to debias adolescents against causal illusions. PLoS One. 2013 Aug 14;8(8):e71303.

Barrett, Deirdre. Supernormal Stimuli: How Primal Urges Overran Their Evolutionary Purpose. New York: W.W. Norton 2010.

Bassler BL, Losick R. Bacterially speaking. Cell. 2006 Apr 21;125(2):237-46.

Beauregard M, O'Leary D. The Spiritual Brain: A Neuroscientist's Case for the Existence of the Soul. HarperOne 2008.

Beck DM, Muggleton N, Walsh V, Lavie N. Right parietal cortex plays a critical role in change blindness. Cereb Cortex. 2006;16(5):712-717.

Beck F. Quantum brain dynamics and consciousness," in The Physical Nature of Consciousness, ed. by P. van Loocke, Amsterdam: Benjamins, pp. 83-116, 2001.

Beckers G, Zeki S. The consequences of inactivating areas V1 and V5 on visual motion perception. Brain. 1995 Feb;118 (Pt 1):49-60.

Bédécarrats A, Chen S, Pearce K, Cai D, Glanzman DL. RNA from trained Aplysia can induce an epigenetic engram forlong-term sensitization in untrained Aplysia. eNeuro. 2018 May 21;5(3). pii: ENEURO.0038-18.2018.

Benjamin P, Crossley M. Gastropod Feeding Systems: Evolution of Neural Circuits that Generate Diverse Behaviors. Oxford Research Encyclopedias. 2020.

Bergagna E, Tartaglia S. Self-Esteem, Social Comparison, and Facebook Use. Eur J Psychol. 2018 Nov 30;14(4):831-845.

Berkowitz A. Expanding our horizons: central pattern generation in the context of complex activity sequences. J Exp Biol. 2019 Oct 15;222(Pt 20). pii: jeb192054.

Bezares-Calderón LA, Berger J, Jasek S, Verasztó C, Mendes S, Gühmann M, Almeda R, Shahidi R, Jékely G. Neural circuitry of a polycystin-mediated hydrodynamic startle response for predator avoidance. Elife. 2018 Dec 14;7. pii: e36262.

Bilewicz M, Winiewski M, Kofta M, Wójcik A. Harmful ideas: the structure and consequences of anti-Semitic beliefs in Poland. Political Psychology. 2013;34:821-839.

Binda P, Morrone MC. Vision During Saccadic Eye Movements. Annu Rev Vis Sci. 2018 Sep 15;4:193-213.

Bloom P. Empathy and Its Discontents. Trends Cogn Sci. 2017 Jan;21(1):24-31.

Bonnefond M, Kastner S, Jensen O. Communication between Brain Areas Based on Nested Oscillations. eNeuro. 2017;4(2):ENEURO.0153-16.2017.

Bor D. The Ravenous Brain: How the new science of consciousness explains our insatiable search for meaning. Basic Books 2012.

Bostrom , N. Superintelligence: Paths, Dangers, Strategies. Oxford University Press 2014.

Bourdeau RW, Lee-Gosselin A, Lakshmanan A, Farhadi A, Kumar SR, Nety SP, Shapiro MG. Acoustic reporter genes for noninvasive imaging of microorganisms in mammalian hosts. Nature. 2018 Jan 3;553(7686):86-90.

Brown, TG. The intrinsic factors in the act of progression in the mammal, Proc Royal Soc Biol Sci. 1911 Vol 84 (572):308-319.

Brunet T, Larson BT, Linden TA, Vermeij MJA, McDonald K, King N. Light-regulated collective contractility in a multicellular choanoflagellate. Science. 2019 Oct 18;366(6463):326-334.

Burkhardt P, Sprecher SG. Evolutionary origin of synapses and neurons - Bridging the gap. Bioessays. 2017 Oct;39(10).

Cajal, SR. Textura del Sistema Nervioso del Hombre y de Los Vertebrados. 1899-1904. Madrid: Imprenta y Libreria de Nicolás Moya.

Cajal, SR. Recuerdos de mi Vida: Historia de mi Labor Cientifica. 1917. Madrid: Imprenta y Libreria de Nicolas Moya.

Cajal SR. Histologie du système nerveux de l'homme et des vertébrés. Paris: Maloine 1909-1911.

Callahan CD, Hinkebein JH. Assessment of anosmia after traumatic brain injury: performance characteristics of the University of Pennsylvania Smell Identification Test. J Head Trauma Rehabil. 2002;17(3):251-256.

Campbell T, Friesen J. Why People "Fly from Facts". Research shows the appeal of untestable beliefs and how they lead to a polarized society. Scientific American, March 3, 2015.

Carey N. Junk DNA: A journey through the dark matter of the genome". Columbia University Press 2015.

Carrillo-Reid L, Han S, Yang W, Akrouh A, Yuste R. Controlling Visually Guided Behavior by Holographic Recalling of Cortical Ensembles. Cell. 2019 Jul 11;178(2):447-457.e5.

Carter OL, Pettigrew JD, Burr DC, Alais D, Hasler F, Vollenweider FX. Psilocybin impairs high-level but not low-level motion perception. Neuroreport. 2004 Aug 26;15(12):1947-51.

Cashin P, Goldsack L, Hall D, O'Toole R. Contrasting signal transduction mechanisms in bacterial and eukaryotic gene transcription. FEMS Microbiol Lett. 2006 Aug;261(2):155-64.

Changizi MA. 'Perceiving the present' as a framework for ecological explanations of the misperception of projected angle and angular size. Perception. 2001;30(2):195-208.

Changizi MA, Zhang Q, Shimojo S. Bare skin, blood and the evolution of primate colour vision. Biol Lett. 2006 Jun 22;2(2):217-21.

Changizi M. The Vision Revolution: How the Latest Research Overturns Everything We Thought We Knew About Human Vision. BenBella Books 2009.

Cohen-Armon M, Visochek L, Katzoff A, Levitan D, Susswein AJ, Klein R, Valbrun M, Schwartz JH. Long-term memory requires polyADP-ribosylation. Science. 2004 Jun 18;304(5678):1820-2.

Colbourne JK et al. The ecoresponsive genome of Daphnia pulex. Science. 2011 Feb 4;331(6017):555-61.

Commissiong JW, Sauve Y. Neurophysiological basis of functional recovery in the neonatal spinalized rat. Exp Brain Res. 1993;96(3):473-9.

Coulon M, Deputte BL, Heyman Y, Baudoin C. Individual recognition in domestic cattle (Bos taurus): evidence from 2D-images of heads from different breeds. PLoS One. 2009;4(2):e4441.

Coviltir V, Burcel M, Cherecheanu AP, Ionescu C, Dascalescu D, Potop V, Burcea M. Update on Myopia Risk Factors and Microenvironmental Changes. J Ophthalmol. 2019 Oct 31;2019:4960852.

Craik KJ. The effect of adaptation on differential brightness discrimination. J Physiol. 1938;92(4):406-421.

Craik KJW, Vernon, MD. The nature of dark adaptation. British Journal of Psychology. General Section. 1941 July;32(1): 62-81.

Craik, KJW. The Nature of Explanation. Cambridge: Cambridge University Press 1943.

Craik, KJW. The Nature of Psychology: A Collection of Papers and Other Writings by the Late Kenneth J W Craik edited by Sherwood. Cambridge, Cambridge University Press 1966.

Crick F. Function of the thalamic reticular complex: the searchlight hypothesis. Proc Natl Acad Sci U S A. 1984 Jul;81(14):4586-90.

Cuthbertson L, Nodwell JR. The TetR family of regulators. Microbiol Mol Biol Rev. 2013 Sep;77(3):440-475.

Danek AH, Öllinger M, Fraps T, Grothe B, Flanagin VL. An fMRI investigation of expectation violation in magic tricks. Front Psychol. 2015 Feb 4;6:84.

Dehaene S, Kerszberg M, Changeux JP. A neuronal model of a global workspace in effortful cognitive tasks. Proc Natl Acad Sci U S A. 1998 Nov 24;95(24):14529-34.

Del Campo J, Ruiz-Trillo I. Environmental survey meta-analysis reveals hidden diversity among unicellular opisthokonts. Mol Biol Evol. 2013 Apr;30(4):802-5.

Del Negro CA, Funk GD, Feldman JL. Breathing matters. Nat Rev Neurosci. 2018;19(6):351-367.

Demertzi A, Tagliazucchi E, Dehaene S, Deco G, Barttfeld P, Raimondo F, Martial C, Fernández-Espejo D, Rohaut B, Voss HU, Schiff ND, Owen AM, Laureys S, Naccache L, Sitt JD. Human consciousness is supported by dynamic complex patterns of brain signal coordination. Sci Adv. 2019 Feb 6;5(2):eaat7603.

Di Domenico SI, Ryan RM. The Emerging Neuroscience of Intrinsic Motivation: A New Frontier in Self-Determination Research. Front Hum Neurosci. 2017 Mar 24;11:145.

Drew T, Võ ML, Wolfe JM. The invisible gorilla strikes again: sustained inattentional blindness in expert observers. Psychol Sci. 2013 Sep;24(9):1848-53.

Dodd MS, Papineau D, Grenne T, Slack JF, Rittner M, Pirajno F, O'Neil J, Little CT. Evidence for early life in Earth's oldest hydrothermal vent precipitates. Nature. 2017 Mar 1;543(7643):60-64.

DoVale E. High Frame Rate Psychophysics: Experimentation to Determine a JND for Frame Rate. SMPTE Motion Imaging Journal. 2017;126(9):41-47.

Dzhambov AM, Dimitrova DD. Exposure-response relationship between traffic noise and the risk of stroke: a systematic review with meta-analysis. Arh Hig Rada Toksikol. 2016 Jun 1;67(2):136-51.

Dzhambov AM, Lercher P. Road Traffic Noise Exposure and Depression/Anxiety: An Updated Systematic Review and Meta-Analysis. Int J Environ Res Public Health. 2019 Oct 27;16(21):4134.

Ebbesen CL, Insanally MN, Kopec CD, Murakami M, Saiki A, Erlich JC. More than Just a "Motor": Recent Surprises from the Frontal Cortex. J Neurosci. 2018;38(44):9402-9413.

Eckler P, Kalyango Y, Paasch E. Facebook use and negative body image among U.S. college women. Women Health. 2017 Feb;57(2):249-267.

Eklund R, Wiens S. Visual awareness negativity is an early neural correlate of awareness: A preregistered study with two Gabor sizes. Cogn Affect Behav Neurosci. 2018 Feb;18(1):176-188.

Elizondo-Garza FJ. Noise problems, savage approaches. From just forget it, to physical violence. The Journal of the Acoustical Society of America. 1999;105 (2):942.

Engen HG, Singer T. Compassion-based emotion regulation up-regulates experienced positive affect and associated neural networks. Soc Cogn Affect Neurosci. 2015 Sep;10(9):1291-301.

Epstein, MJ. Sound and Noise: A Listener's Guide to Everyday Life. McGill-Queen's University Press 2020.

Evans PD, Gilbert SL, Mekel-Bobrov N, et al. Microcephalin, a gene regulating brain size, continues to evolve adaptively in humans. Science. 2005;309(5741):1717-1720.

Ezkurdia I, Juan D, Rodriguez JM, Frankish A, Diekhans M, Harrow J, Vazquez J, Valencia A, Tress ML. Multiple evidence strands suggest that there may be as few as 19,000 human protein-coding genes. Hum Mol Genet. 2014 Nov 15;23(22):5866-78.

Ferreira FR, Nogueira MI, Defelipe J. The influence of James and Darwin on Cajal and his research into the neuron theory and evolution of the nervous system. Front Neuroanat. 2014 Jan 29;8:1.

Feinberg TE, Mallatt JM. The Ancient Origins of Consciousness: How the Brain Created Experience Paperback. The MIT Press 2016.

Fiore VG, Dolan RJ, Strausfeld NJ, Hirth F. Evolutionarily conserved mechanisms for the selection and maintenance of behavioural activity. Philos Trans R Soc Lond B Biol Sci. 2015 Dec 19;370(1684):20150053.

Fishell G, Kepecs A. Interneuron Types as Attractors and Controllers. Annu Rev Neurosci. 2019 Jul 12.

Ford CL, Wallace SP, Newman SP, Lee SJ, Cuningham WE. Belief in AIDS-related conspiracy theories and mistrust in the government: relationship with HIV testing among at-risk older adults. The Gerontologist. 2013;53, 973-984.

Francis D. On the eyespot of the dinoflagellate, Nematodinium. J Exp Biol. 1967 Dec;47(3):495-501.

Franklin DW, Reichenbach A, Franklin S, Diedrichsen J. Temporal Evolution of Spatial Computations for Visuomotor Control. J Neurosci. 2016 Feb 24;36(8):2329-41.

Fraser, J. (1908). A new visual illusion of direction. British Journal of Psychology, 2, 307-320.

Frith CD. The neural basis of consciousness. Psychol Med. 2019;1-13.

Gallotto S, Sack AT, Schuhmann T, de Graaf TA. Oscillatory Correlates of Visual Consciousness. Front Psychol. 2017;8:1147.

Gallup GH, Newport T. Belief in paranormal phenomena among adult Americans. The Skeptical Inquirer. 1991;15, 137-146.

Galperin MY. What bacteria want. Environ Microbiol. 2018 Dec;20(12):4221-4229. doi: 10.1111/1462-2920.14398. Epub 2018 Oct 25.

Gavelis GS, Hayakawa S, White RA 3rd, Gojobori T, Suttle CA, Keeling PJ, Leander BS. Eye-like ocelloids are built from different endosymbiotically acquired components. Nature. 2015 Jul 9;523(7559):204-7.

Gazzaniga MS. The Consciousness Instinct: Unraveling the Mystery of How the Brain Makes the Mind. Farrar, Straus and Giroux 2019.

Gelderblom H and Menge L. The invisible gorilla revisited: using eye tracking to investigate inattentional blindness in interface design. AVI '18: Proceedings of the 2018 International Conference on Advanced Visual Interfaces. May 2018 Article No. 39.

Ghahramani Z. Probabilistic machine learning and artificial intelligence. Nature. 2015 May 28;521(7553):452-9.

Gindrat AD, Chytiris M, Balerna M, Rouiller EM, Ghosh A. Use-dependent cortical processing from fingertips in touchscreen phone users. Curr Biol. 2015 Jan 5;25(1):109-16.

Glennon E, Carcea I, Martins ARO, Multani J, Shehu I, Svirsky MA, Froemke RC. Locus coeruleus activation accelerates perceptual learning. Brain Res. 2019 Apr 15;1709:39-49.

Goetz JL, Keltner D, Simon-Thomas E. Compassion: an evolutionary analysis and empirical review. Psychol Bull. 2010 May;136(3):351-74.

Goldberg S, Visochek L, Giladi E, Gozes I, Cohen-Armon M. PolyADP-ribosylation is required for long-term memory formation in mammals. J Neurochem. 2009 Oct;111(1):72-9.

Golgi C. Sulla struttura della sostanza grigia del cervello (Comunicazione preventiva). Gaz. Med. Ital. Lomb. 1873;33, 244-246.

Gordon G, Ahissar E. Hierarchical curiosity loops and active sensing. Neural Netw. 2012 Aug;32:119-29.

Gottschalk MG, Domschke K. Genetics of generalized anxiety disorder and related traits. Dialogues Clin Neurosci. 2017;19(2):159-168.

Grillner S, El Manira A. Current Principles of Motor Control, with Special Reference to Vertebrate Locomotion. Physiol Rev. 2020 Jan 1;100(1):271-320.

Gruber MJ, Ranganath C. How Curiosity Enhances Hippocampus-Dependent Memory: The Prediction, Appraisal, Curiosity, and Exploration (PACE) Framework. Trends Cogn Sci. 2019 Dec;23(12):1014-1025.

Guillery RW, Sherman SM. Thalamic relay functions and their role in corticocortical communication: generalizations from the visual system. Neuron. 2002 Jan 17;33(2):163-75.

Gupta PD, Muthukumar A. Minor to Chronic Eye Disorders Due to Environmental Pollution: A Review. J Ocul Infect Inflamm 2018,2:2

Harris S, Sheth SA, Cohen MS. Functional neuroimaging of belief, disbelief, and uncertainty. Ann Neurol. 2008;63(2):141-147.

Hart AC, Sims S, Kaplan JM. Synaptic code for sensory modalities revealed by C. elegans GLR-1 glutamate receptor. Nature. 1995 Nov 2;378(6552):82-5.

Higgins E, Rayner K. Transsaccadic processing: stability, integration, and the potential role of remapping. Atten Percept Psychophys. 2015 Jan;77(1):3-27.

Hoffman DD, Singh M. Computational evolutionary perception. Perception. 2012;41(9):1073-91.

Horridge GA. Presumed photoreceptive cilia in a ctenophore. Quart J Microsc Sci. 1964;105:311-7.

Huang HM, Chang DS, Wu PC. The Association between Near Work Activities and Myopia in Children-A Systematic Review and Meta-Analysis. PLoS One. 2015 Oct 20;10(10):e0140419.

Hublin JJ, Ben-Ncer A, Bailey SE, et al. New fossils from Jebel Irhoud, Morocco and the pan-African origin of *Homo sapiens* [published correction appears in Nature. 2018 Jun;558(7711):E6]. Nature. 2017;546(7657):289-292.

Huckins JF, daSilva AW, Wang R, et al. Fusing Mobile Phone Sensing and Brain Imaging to Assess Depression in College Students. Front Neurosci. 2019 Mar 21;13:248

Jasun Horsley, Dark Oasis: A Self-Made Messiah Unveiled, Auticulture, 2017.

Johnston I. 'Mismatch' between the way our senses evolved and modern world is making us ill, experts warn. The Independent, Monday 20 February 2017.

Haggbloom SJ, Warnick R, Warnick JE et al. . The 100 most eminent psychologists of the 20th century. Review of General Psychology. 2002;6, 139-152.

Hahad O, Prochaska JH, Daiber A, Muenzel T. Environmental Noise-Induced Effects on Stress Hormones, Oxidative Stress, and Vascular Dysfunction: Key Factors in the Relationship between Cerebrocardiovascular and Psychological Disorders. Oxid Med Cell Longev. 2019 Nov 11;2019:4623109.

Hammer MS, Swinburn TK, Neitzel RL. Environmental noise pollution in the United States: developing an effective public health response. Environmental Health Perspectives. 2014;122 (2): 115-9.

Harraldsson E, Houtkooper J. Traditional Christian beliefs, spiritualism, and the paranormal: an Icelandic-American comparison. The International Journal for the Psychology of Religion. 1996;6, 51-64.

Harris S. Free Will. Free Press 2012.

Hausler M, Strecker C, Huber A, Brenner M, Höge T, Höfer S. Distinguishing Relational Aspects of Character Strengths with Subjective and Psychological Well-being. Front Psychol. 2017 Jul 11;8:1159.

Hergovich A, Schott R, Arendasy M. On the relationship between paranormal belief and schizotypy among adolescents. Personality and Individual Differences. 2008:45, 119-125.

Hiramatsu C, Melin AD, Allen WL, Dubuc C, Higham JP. Experimental evidence that primate trichromacy is well suited for detecting primate social colour signals. Proc Biol Sci. 2017 Jun 14;284(1856):20162458.

Hofmann SG, Grossman P, Hinton DE. Loving-kindness and compassion meditation: potential for psychological interventions. Clin Psychol Rev. 2011 Nov;31(7):1126-32.

Hosoya H, Hyvärinen A. A mixture of sparse coding models explaining properties of face neurons related to holistic and parts-based processing. PLoS Comput Biol. 2017 Jul 25;13(7):e1005667.

Hsieh YJ, Wanner BL. Global regulation by the seven-component Pi signaling system. Curr Opin Microbiol. 2010 Apr;13(2):198-203.

Huda R, Goard MJ, Pho GN, Sur M. Neural mechanisms of sensorimotor transformation and action selection. Eur J Neurosci. 2019 Apr;49(8):1055-1060.

Hussain Z, Griffiths MD. Problematic Social Networking Site Use and Comorbid Psychiatric Disorders: A Systematic Review of Recent Large-Scale Studies. Front Psychiatry. 2018 Dec 14;9:686.

Ihara K, Umemura T, Katagiri I, Kitajima-Ihara T, Sugiyama Y, Kimura Y, Mukohata Y. Evolution of the archaeal rhodopsins: evolution rate changes by gene duplication and functional differentiation. J Mol Biol. 1999 Jan 8;285(1):163-74.

Jafek BW, Linschoten MR, Murrow BW. 2004. Anosmia after intranasal zinc gluconate use. Am J Rhinol 18(3):137-141.

Jékely G, Colombelli J, Hausen H, Guy K, Stelzer E, Nédélec F, Arendt D. Mechanism of phototaxis in marine zooplankton. Nature. 2008 Nov 20;456(7220):395-9.

Jékely G, Melzer S, Beets I, Kadow ICG, Koene J, Haddad S, Holden-Dye L. The long and the short of it - a perspective on peptidergic regulation of circuits and behaviour. J Exp Biol. 2018 Feb 8;221(Pt 3).

Jolley D, Douglas, KM. The social consequences of conspiracism: exposure to conspiracy theories decreases intentions to engage in politics and to reduce one's carbon footprint. British Journal of Psychology. 2014;105, 35-56.

Jones L, Sciamanna C, Lehman E. Are those who use specific complementary and alternative medicine therapies less likely to be immunized? Preventive Medicine. 2010;50, 148-154.

Joosse, P. Silence, Charisma and Power: The Case of John de Ruiter. Journal of Contemporary Religion. 2006;21(3) 355-371.

Jovanovic V, Brdaric D. Did curiosity kill the cat? Evidence from subjective well-being in adolescents. Personal. Indiv. Diff. 2012;52 (3): 380-384.

Kashdan TB, Elhai JD, Breen WE. Social anxiety and disinhibition: an analysis of curiosity and social rank appraisals, approach-avoidance conflicts, and disruptive risk-taking behavior. J Anxiety Disord. 2008 Aug;22(6):925-39.

Katz PS. Evolution of central pattern generators and rhythmic behaviours. Philos Trans R Soc Lond B Biol Sci. 2016 Jan 5;371(1685):20150057.

Karten HJ. The organization of the avian telencephalon and some speculations on the phylogeny of the amniote telencephalon. Annals of the New York Academy of Sciences. vol 67(1):164-179.

Kean S. The violinist's thumb: other tales of love, war, and genius, as written by our genetic code. Little Brown Company, Hachette Book Group, 2014.

Keijzer F, van Duijn M, Lyon P. What nervous systems do: early evolution, input-output, and the skin brain thesis. Adaptive Behavior 2013, 21(2):67-85.

Kelley DB, Ballagh IH, Barkan CL, Bendesky A, Elliott TM, Evans BJ, Hall IC, Kwon YM, Kwong-Brown U, Leininger EC, Perez EC, Rhodes HJ, Villain A, Yamaguchi A, Zornik E.

Generation, Coordination, and Evolution of Neural Circuits for Vocal Communication. J Neurosci. 2020 Jan 2;40(1):22-36.

Kelmer Sacramento E, Kirkpatrick JM, Mazzetto M, et al. Reduced proteasome activity in the aging brain results in ribosome stoichiometry loss and aggregation. Mol Syst Biol. 2020;16(6):e9596.

Kim JN, Shadlen MN. Neural correlates of a decision in the dorsolateral prefrontal cortex of the macaque. Nat Neurosci. 1999 Feb;2(2):176-85.

Knolle F, Goncalves RP, Morton AJ. Sheep recognize familiar and unfamiliar human faces from two-dimensional images. R Soc Open Sci. 2017 Nov 8;4(11):171228.

Khosravipour M, Khanlari P. The association between road traffic noise and myocardial infarction: A systematic review and meta-analysis. Sci Total Environ. 2020 Aug 20;731:139226.

Klimecki OM, Leiberg S, Lamm C, Singer T. Functional neural plasticity and associated changes in positive affect after compassion training. Cereb Cortex. 2013 Jul;23(7):1552-61.

Kofoid CA, Swezy O. The free-living, unarmoured Dinoflagellates. Mem Univ Calif. 1921;5:1-562.

Kohda M, Jordan LA, Hotta T, Kosaka N, Karino K, Tanaka H, Taniyama M, Takeyama T. Facial Recognition in a Group-Living Cichlid Fish. PLoS One. 2015 Nov 25;10(11):e0142552.

Krummenacher P, Mohr C, Haker H, Brugger P. Dopamine, paranormal belief, and the detection of meaningful stimuli. J Cogn Neurosci. 2010;22(8):1670-1681.

Kuriki I, Ashida H, Murakami I, Kitaoka A. Functional brain imaging of the Rotating Snakes illusion by fMRI. J Vis. 2008 Dec 30;8(10):16.1-10.

Lafreniere D, Mann N. Anosmia: loss of smell in the elderly. Otolaryngol Clin North Am. 2009;42(1):123-131.

Lanier , JZ. Ten Arguments for Deleting Your Social Media Accounts Right Now, Henry Holt and Co. 2018.

Le Floch A, Ropars G. Left-right asymmetry of the Maxwell spot centroids in adults without and with dyslexia. Proc Biol Sci. 2017 Oct 25;284(1865). pii: 20171380.

Lee AT, Cunniff MM, See JZ, Wilke SA, Luongo FJ, Ellwood IT, Ponnavolu S, Sohal VS. VIP Interneurons Contribute to Avoidance Behavior by Regulating Information Flow across Hippocampal-Prefrontal Networks. Neuron. 2019 Jun 19;102(6):1223-1234.e4.

Lee H, Choi MK, Lee D, Kim HS, Hwang H, Kim H, Park S, Paik YK, Lee J. Nictation, a dispersal behavior of the nematode Caenorhabditis elegans, is regulated by IL2 neurons. Nat Neurosci. 2011 Nov 13;15(1):107-12.

Leitch IJ. Genome sizes through the ages. Heredity. Nature Publishing Group. 2007; 99 (2): 121-122.

Leys SP, Mah JL, McGill PR, Hamonic L, De Leo FC, Kahn AS. Sponge Behavior and the Chemical Basis of Responses: A Post-Genomic View. Integr Comp Biol. 2019 Oct 1;59(4):751-764.

Libet B, Gleason CA, Wright EW, Pearl DK. Time of conscious intention to act in relation to onset of cerebral activity (readiness-potential). The unconscious initiation of a freely voluntary act. Brain. 1983 Sep;106 (Pt 3):623-42.

Lin W, Kirschvink JL, Paterson GA, Bazylinski DA, Pan Y. On the origin of microbial magnetoreception. National Science Review, 2019, nwz065.

Liu H, Chaudhury D. Understanding Mood Disorders Using Electrophysiology and Circuit Breaking. In: Çelik A., Wernet M. (eds) Decoding Neural Circuit Structure and Function. Springer, Cham 2017.

Loidolt M, Aust U, Meran I, Huber L. Pigeons use item-specific and category-level information in the identification and categorization of human faces. J Exp Psychol Anim Behav Process. 2003 Oct;29(4):261-76.

Lugaro E. I recenti progressi dell' anatomia del sistema nervoso in rapporto alla psicologia ed alla psichiatria. Riu. d. Palol. neru. e mentale., t. IV, fasc. 11-12, 1891

Luo S, Xu H, Zuo Y, Liu X, All AH. A Review of Functional Electrical Stimulation Treatment in Spinal Cord Injury. Neuromolecular Med. 2020 Jan 8.

Lydon-Staley DM, Zurn P, Bassett DS. Within-person variability in curiosity during daily life and associations with well-being. J Pers. 2020 Aug;88(4):625-641.

Lyon P. The cognitive cell: bacterial behavior reconsidered. Front Microbiol. 2015 Apr 14;6:264.

Macaskill A and Denovan A. Assessing psychological health: the contribution of psychological strengths, British Journal of Guidance & Counselling. 2014;42:3, 320-337.

Maddocks SE, Oyston PCF. Structure and function of the LysR-type transcriptional regulator (LTTR) family proteins. Microbiology. 2008 Dec;154(Pt 12):3609-3623.

Magalhaes AA, Oliveira L, Pereira MG, Menezes CB. Does Meditation Alter Brain Responses to Negative Stimuli? A Systematic Review. Front Hum Neurosci. 2018;12:448.

Margulis B, Tsimokha A, Zubova S, Guzhova I. Molecular Chaperones and Proteolytic Machineries Regulate Protein Homeostasis In Aging Cells. Cells. 2020;9(5):1308.

Marín O. Interneuron dysfunction in psychiatric disorders. Nat Rev Neurosci. 2012 Jan 18;13(2):107-20.

Martin W, Russell MJ. On the origins of cells: a hypothesis for the evolutionary transitions from abiotic geochemistry to chemoautotrophic prokaryotes, and from prokaryotes to nucleated cells. Philos Trans R Soc Lond B Biol Sci. 2003 Jan 29;358(1429):59-83.

Mascaro JS, Rilling JK, Tenzin Negi L, Raison CL. Compassion meditation enhances empathic accuracy and related neural activity. Soc Cogn Affect Neurosci. 2013 Jan;8(1):48-55.

McLeish JD. Children's superstitions: British and Canadian. Canadian Journal of Education. 1984;19, 425-436.

McCarthy D, Lueras P, Bhide PG. Elevated dopamine levels during gestation produce region-specific decreases in neurogenesis and subtle deficits in neuronal numbers. Brain Res. 2007; 1182: 11-25.

McCrann S, Loughman J, Butler JS, Paudel N, Flitcroft DI. Smartphone use as a possible risk factor for myopia. Clin Exp Optom. 2020 May 25.

Mekel-Bobrov N, Gilbert SL, Evans PD, et al. Ongoing adaptive evolution of ASPM, a brain size determinant in *Homo sapiens.* Science. 2005;309(5741):1720-1722.

Melcher D, Colby CL. Trans-saccadic perception. Trends Cogn Sci. 2008 Dec;12(12):466-73.

Mély DA, Linsley D, Serre T. Complementary surrounds explain diverse contextual phenomena across visual modalities. Psychol Rev. 2018 Oct;125(5):769-784.

Messias E, Kindrick K, Castro J. School bullying, cyberbullying, or both: correlates of teen suicidality in the 2011 CDC Youth Risk Behavior Survey. Compr Psychiatry. 2014 Jul;55(5):1063-8.

Miller SE, Legan AW, Henshaw MT, Ostevik KL, Samuk K, Uy FMK, Sheehan MJ. Evolutionary dynamics of recent selection on cognitive abilities. Proc Natl Acad Sci U S A. 2020 Feb 11;117(6):3045-3052.

Miyata M, et al. Tree of motility - A proposed history of motility systems in the tree of life. Genes Cells. 2020 Jan;25(1):6-21.

Modolo J, Hassan M, Wendling F, Benquet P. Decoding the circuitry of consciousness: From local microcircuits to brain-scale networks. Netw Neurosci. 2020;4(2):315-337.

Molkov YI, Shevtsova NA, Park C, Ben-Tal A, Smith JC, Rubin JE, Rybak IA. A closed-loop model of the respiratory system: focus on hypercapnia and active expiration. PLoS One. 2014 Oct 10;9(10):e109894.

Mongillo P, Scandurra A, Kramer RSS, Marinelli L. Recognition of human faces by dogs (Canis familiaris) requires visibility of head contour. Anim Cogn. 2017 Sep;20(5):881-890.

Moran NA. Microbial minimalism: genome reduction in bacterial pathogens. Cell. 2002 Mar 8;108(5):583-6.

Mugaloglu, EZ. The Problem of Pseudoscience in Science Education and Implications of Constructivist Pedagogy. Science & Education, v23 n4 p829-842 Apr 2014.

Mukamel R, Ekstrom AD, Kaplan J, Iacoboni M, Fried I. Single-neuron responses in humans during execution and observation of actions. Curr Biol. 2010 Apr 27;20(8):750-6.

Münzel T, Kröller-Schön , Oelze M. Adverse Cardiovascular Effects of Traffic Noise with a Focus on Nighttime Noise and the New WHO Noise Guideline. Annual Review of Public Health. 2020;41: 309-328.

Muzaffar N, Brito EB, Fogel J, Fagan D, Kumar K, Verma R. The Association of Adolescent Facebook Behaviours with Symptoms of Social Anxiety, Generalized Anxiety, and Depression. J Can Acad Child Adolesc Psychiatry. 2018 Nov;27(4):252-260. Epub 2018 Nov 1.

National Science Board (2006), figure 7-8 - Belief in paranormal phenomena: 1990, 2001, and 2005. "figure 7-8". David W. Moore (16 June 2005). "Three in Four Americans Believe in Paranormal". "Appendix table 7-16: Attitudes toward science and technology, by country/region: Most recent year". Science and Engineering Indicators 2006. Volume 2: Appendix Tables. Arlington, VA: National Science Foundation. pp. A7-17.

Nahin RL, Barnes, PM, Stussman BJ, Bloom B. Costs of complimentary and alternative medicine (CAM) and frequency of visits to CAM practitioners: United States 2007. National Health Statistics Reports. 2009;18, 1-14.

Newberg A, Waldman MR. How Enlightenment Changes Your Brain: The New Science of Transformation. Avery 2016.

Newby RW, Davis JB. Relationships between locus of control and paranormal beliefs. Psychological Reports. 2004;94, 1261-1266.

Nordström K, Wallén R, Seymour J, Nilsson D. A simple visual system without neurons in jellyfish larvae. Proc Biol Sci. 2003 Nov 22;270(1531):2349-54.

Oliver JE, Wood TJ. Conspiracy theories and the paranoid style(s) of mass opinion. American Journal of Political Science. 2014;58, 952-966.

Orenstein, A. Religion and paranormal belief. Journal for the Scientific Study of Religion. 2002;41, 301-311.

Oudeyer PY, Gottlieb J, Lopes M. Intrinsic motivation, curiosity, and learning: Theory and applications in educational technologies. Prog Brain Res. 2016;229:257-284.

Pace TW, Negi LT, Adame DD, Cole SP, Sivilli TI, Brown TD, Issa MJ, Raison CL. Effect of compassion meditation on neuroendocrine, innate immune and behavioral responses to psychosocial stress. Psychoneuroendocrinology. 2009 Jan;34(1):87-98.

Pal D, Dean JG, Liu T, Li D, Watson CJ, Hudetz AG, Mashour GA. Differential Role of Prefrontal and Parietal Cortices in Controlling Level of Consciousness. Curr Biol. 2018 Jul 9;28(13):2145-2152.e5.

Pardo L, Ballesteros JA, Osman R, Weinstein H. On the use of the transmembrane domain of bacteriorhodopsin as a template for modeling the three-dimensional structure of guanine nucleotide-binding regulatory protein-coupled receptors. Proc Natl Acad Sci U S A. 1992 May 1;89(9):4009-12.

Parris BA, Kuhn G, Mizon GA, Benattayallah A, Hodgson TL. Imaging the impossible: an fMRI study of impossible causal relationships in magic tricks. Neuroimage. 2009 Apr 15;45(3):1033-9.

Pierce AH. The illusion of the kindergarten patterns. Psychological Review. 1898;5(3), 233-253.

Popple AV, Sagi D. A Fraser illusion without local cues? Vision Res. 2000;40(8):873-8.

Polyak SL. The retina. Chicago: University of Chicago Press; 1941.

Proulx MJ, Ptito M, Amedi A. Multisensory integration, sensory substitution and visual rehabilitation. Neurosci Biobehav Rev. 2014 Apr;41:1-2.

Quiroga RQ. Magic and cognitive neuroscience. Curr Biol. 2016 May 23;26(10):R390-4.

Raichle ME. The brain's default mode network. Annu Rev Neurosci. 2015 Jul 8;38:433-47.

Ramírez-Barrantes R, Arancibia M, Stojanova J, Aspé-Sánchez M, Córdova C, Henríquez-Ch RA. Default Mode Network, Meditation, and Age-Associated Brain Changes: What Can We Learn from the Impact of Mental Training on Well-Being as a Psychotherapeutic Approach? Neural Plast. 2019 Apr 2;2019:7067592.

Rice TW. Believe it or not: religious and other paranormal beliefs in the United States. Journal for the Scientific Study of Religion. 2003;42, 95-106.

Rieiro H, Martinez-Conde S, Macknik SL. Perceptual elements in Penn & Teller's "Cups and Balls" magic trick. PeerJ. 2013 Feb 12;1:e19.

Ritchie H, Roser M Urbanization. 2018. Published online at OurWorldInData.org. Retrieved from: 'https://ourworldindata.org/urbanization.

Rizzolatti G, Fadiga L, Gallese V, Fogassi L. Premotor cortex and the recognition of motor actions. Brain Res Cogn Brain Res. 1996 Mar;3(2):131-41.

Rokszin A, Márkus Z, Braunitzer G, Berényi A, Benedek G, Nagy A. Visual athways serving motion detection in the mammalian brain. Sensors (Basel). 2010;10(4):3218-42.

Rossini PM, Di Iorio R, Bentivoglio M, Bertini G, Ferreri F, Gerloff C, Ilmoniemi RJ, Miraglia F, Nitsche MA, Pestilli F, Rosanova M, Shirota Y, Tesoriero C, Ugawa Y, Vecchio F, Ziemann U, Hallett M. Methods for analysis of brain connectivity: An IFCN-sponsored review. Clin Neurophysiol. 2019 Oct;130(10):1833-1858.

Ryan K, Lu Z, Meinertzhagen IA. The CNS connectome of a tadpole larva of Ciona intestinalis (L.) highlights sidedness in the brain of a chordate sibling. Elife. 2016 Dec 6;5. pii: e16962.

Ryan TJ, Brown J, Johnson A, Sanberg C, Schildmier M. Science literacy and belief in the paranormal-an empirical test. Skeptic. 2004;10, 12-13.

Sagone E, De Caroli ME. Locus of control and beliefs about superstition and luck in adolescents: what's their relationship? Procedia - Social and Behavioral Sciences. 2014;140, 318-323.

Sato W, Kochiyama T, Uono S. Spatiotemporal neural network dynamics for the processing of dynamic facial expressions. Sci Rep. 2015 Jul 24;5:12432.

Sauerbrei BA, Guo JZ, Cohen JD, Mischiati M, Guo W, Kabra M, Verma N, Mensh B, Branson K, Hantman AW. Cortical pattern generation during dexterous movement is input-driven. Nature. 2020 Jan;577(7790):386-391.

Scales D, Gorman S, Gorman J. Resist pseudoscience with respect, not ridicule. Nature. 2020 Jun;582(7810):32.

Schmitgen MM, Horvath J, Mundinger C, Wolf ND, Sambataro F, Hirjak D, Kubera KM, Koenig J, Wolf RC. Neural correlates of cue reactivity in individuals with smartphone addiction. Addict Behav. 2020 Sep;108:106422.

Schnitzler C, Pang K, Powers M, Reitzel A, Ryan J, Simmons D, et al. Genomic organization, evolution, and expression of photoprotein and opsin genes in Mnemiopsis leidyi: a new view of ctenophore photocytes. BMC Biol. 2012;10:107.

Sebé-Pedrós A, Degnan BM, Ruiz-Trillo I. The origin of Metazoa: a unicellular perspective. Nat Rev Genet. 2017 Aug;18(8):498-512.

Seth C. 30 Second Brain. Sterling; Later Printing Edition (January 1, 2014).

Sevinc G, Greenberg J, Hölzel BK, et al. Hippocampal circuits underlie improvements in self-reported anxiety following mindfulness training [published online ahead of print, 2020 Jul 23]. Brain Behav. 2020;e01766.

Shadlen MN, Newsome WT. Neural basis of a perceptual decision in the parietal cortex (area LIP) of the rhesus monkey. J Neurophysiol. 2001 Oct;86(4):1916-36.

Shelchkova N, Tang C, Poletti M. Task-driven visual exploration at the foveal scale. Proc Natl Acad Sci U S A. 2019 Mar 19;116(12):5811-5818.

Sherwin JC, Reacher MH, Keogh RH, Khawaja AP, Mackey DA, Foster PJ. The association between time spent outdoors and myopia in children and adolescents: a systematic review and meta-analysis. Ophthalmology 2012; 119(10): 2141-2151.

Shrestha S, Kamel F, Umbach DM, Freeman LEB, Koutros S, Alavanja M, et al. 2019. High pesticide exposure events and olfactory impairment among U.S. farmers. Environ Health Perspect 127(1):17005,

Siebeck UE, Parker AN, Sprenger D, Mäthger LM, Wallis G. A species of reef fish that uses ultraviolet patterns for covert face recognition. Curr Biol. 2010 Mar 9;20(5):407-10.

Simons DJ, Chabris CF. Gorillas in our midst: Sustained inattentional blindness for dynamic events. Perception. 1999;28:1059-1074.

Sinigaglia C, Busengdal H, Lerner A, Oliveri P, Rentzsch F. Molecular characterization of the apical organ of the anthozoan Nematostella vectensis. Dev Biol. 2015 Feb 1;398(1):120-33.

Stark E. Enhancing and assessing critical thinking in a psychological research methods course. Teaching of Psychology. 2012;39, 107-112.

Steenwinkel FA, Ter Meulen BC, Ramachandran VS. Teaching Video NeuroImages: Mirror therapy against phantom pain. Neurology. 2019;92(9):e1002.

Strauss C, Lever Taylor B, Gu J, Kuyken W, Baer R, Jones F, Cavanagh K. What is compassion and how can we measure it? A review of definitions and measures. Clin Psychol Rev. 2016 Jul;47:15-27.

Strother PK, Battison L, Brasier MD, Wellman CH. Earth's earliest non-marine eukaryotes. Nature. 2011 May 26;473(7348):505-9.

Summers JJ, Anson JG. Current status of the motor program: revisited. Hum Mov Sci. 2009;28(5):566-577.

Suryanarayana SM, Pérez-Fernández J, Robertson B, Grillner S. The evolutionary origin of visual and somatosensory representation in the vertebrate pallium. Nat Ecol Evol. 2020 Mar 16.

Suzuki D, Irieda H, Homma M, Kawagishi I, Sudo Y. Phototactic and chemotactic signal transduction by transmembrane receptors and transducers in microorganisms. Sensors (Basel). 2010;10(4):4010-39.

Swami V. Social psychological origins of conspiracy theories: the case of the Jewish conspiracy theory in Malaysia. Frontiers in Psycholog. 2012;3, 280.

Swan GE, Carmelli D. Curiosity and mortality in aging adults: a 5-year follow-up of the Western Collaborative Group Study. Psychol Aging. 1996 Sep;11(3):449-53.

Takeuchi T. The effect of eccentricity and the adapting level on the café wall illusion. Percept Psychophys. 2005 Oct;67(7):1113-27.

Tang Q, Pang K, Yuan X, Xiao S. A one-billion-year-old multicellular chlorophyte. Nat Ecol Evol. 2020 Feb 24.

The Illusion Index (https://www.illusionsindex.org). Last accessed Sept 21, 2020.

Thoen HH, How MJ, Chiou TH, Marshall J. A different form of color vision in mantis shrimp. Science. 2014 Jan 24;343(6169):411-3.

Thomson EM. Air Pollution, Stress, and Allostatic Load: Linking Systemic and Central Nervous System Impacts. J Alzheimers Dis. 2019;69(3):597-614.

Tobacyk J, Pirttilä-Backman AM. Paranormal beliefs and their implications in university students from Finland and the United States. Journal of Cross-Cultural Psychology. 1992;23, 59-76.

Kashdan TB, Roberts JE. Trait and State Curiosity in the Genesis of Intimacy: Differentiation From Related Constructs. Journal of Social and Clinical Psychology. 2004;23 (6):792-816.

Tolman EC, Honzik CH. Introduction and removal of reward, and maze performance in rats. University of California Publications in Psychology 1930.

Torres MN, Barberia I, Rodríguez-Ferreiro J. Causal illusion as a cognitive basis of pseudoscientific beliefs. Br J Psychol. 2020 Feb 10. doi: 10.1111/bjop.12441. Epub ahead of print.

Tostevin R, Clarkson MO, Gangl S, Shields GA, Wood RA, Bowyer F, ... Stirling CH. Uranium isotope evidence for an expansion of anoxia in terminal Ediacaran oceans. Earth and Planetary Science Letters. 2019;506:104-112.

Ulrich LE, Koonin EV, Zhulin IB. One-component systems dominate signal transduction in prokaryotes. Trends Microbiol. 2005 Feb;13(2):52-6.

Utinans A, Ancane G. Belief in the paranormal and modern health worries. SHS Web of Conferences. 2014;10, 00048.

Vaina LM, Solomon J, Chowdhury S, Sinha P, Belliveau JW. Functional neuroanatomy of biological motion perception in humans. Proc Natl Acad Sci U S A. 2001 Sep 25;98(20):11656-61.

Villmoare B, Kimbel WH, Seyoum C, et al. Paleoanthropology. Early Homo at 2.8 Ma from Ledi-Geraru, Afar, Ethiopia. Science. 2015;347(6228):1352-1355.

Visconti di Oleggio Castello M, Gobbini MI. Familiar Face Detection in 180 ms. PLoS One. 2015 Aug 25;10(8):e0136548.

von Stumm S, Hell B, Chamorro-Premuzic T. The Hungry Mind: Intellectual Curiosity Is the Third Pillar of Academic Performance. Perspect Psychol Sci. 2011 Nov;6(6):574-88.

Walbolt J, Koh Y. Non-nutritive Sweeteners and Their Associations with Obesity and Type 2 Diabetes. J Obes Metab Syndr. 2020 Jun 30;29(2):114-123.

Walker WR, Hoekstra SJ, Vogl RJ. Science education is no guarantee of skepticism. Skeptic. 2002;9:24-28.

Wang Z, Zhao L, Huang Q, Hong A, Yu C, Xiao Q, Zou B, Ji S, Zhang L, Zou K, Ning Y, Zhang J, Jia P. Traffic-related environmental factors and childhood obesity: A systematic review and meta-analysis. Obes Rev. 2020 Jan 30.

Watts KJ, Vaknin A, Fuqua C, Kazmierczak BI. New Twists and Turns in Bacterial Locomotion and Signal Transduction. J Bacteriol. 2019 Jul 29. pii: JB.00439-19.

Wenger, DM. The Illusion of Conscious Will. MIT Press 2003.

White JG, Southgate E, Thomson JN, Brenner S. The structure of the nervous system of the nematode Caenorhabditis elegans.

Philos Trans R Soc Lond B Biol Sci. 1986 Nov 12;314(1165):1-340.

Wiersma CA, Ikeda K. Interneurons commanding swimmeret movements in the Crayfish, *Procambarus clarki (girard)*. Comp Biochem Physiol. 1964 Aug;12:509-25.

Williams DG. The Mind as a Predictive Modelling Engine: Generative Models, Structural Similarity, and Mental Representation (Doctoral thesis). https://doi.org/10.17863/CAM.33386.

Williams DG. Hierarchical Bayesian models of delusion. Conscious Cogn. 2018 May;61:129-147.

Wilson JA. Reducing Pseudoscientific and Paranormal Beliefs in University Students Through a Course in Science and Critical Thinking. Science & Education. 2018;27:183.

Wong PC, Chandrasekaran B, Zheng J. The derived allele of ASPM is associated with lexical tone perception. PLoS One. 2012;7(4):e34243.

Xenakis SN. Neuropsychiatric evidence of waterboarding and other abusive treatments. Torture. 2012;22 Suppl 1:21-4.

Yackle K, Schwarz LA, Kam K, et al. Breathing control center neurons that promote arousal in mice. Science. 2017;355(6332):1411-1415.

Yunker AG, Patel R, Page KA. Effects of Non-nutritive Sweeteners on Sweet Taste Processing and Neuroendocrine Regulation of Eating Behavior. Curr Nutr Rep. 2020 Sep;9(3):278-289.

Zamoscik VE, Schmidt SNL, Gerchen MF, et al. Respiration pattern variability and related default mode network connectivity are altered in remitted depression. Psychol Med. 2018;48(14):2364-2374.

Zamoscik V, Schmidt SNL, Timm C, Kuehner C, Kirsch P. Modulation of respiration pattern variability and its relation to anxiety symptoms in remitted recurrent depression. Heliyon. 2020;6(7):e04261.

Zangwill O. Kenneth Craik: The man and his work. British Journal Of Psychology. 1980;71(1):1-16.

Zaretskaya N, Thielscher A, Logothetis NK, Bartels A. Disrupting parietal function prolongs dominance durations in binocular rivalry. Curr Biol. 2010;20(23):2106-2111.

Zeki S. Cerebral akinetopsia (visual motion blindness). A review. Brain. 1991;114(Pt 2):811-824.

Zelano C, Jiang H, Zhou G, et al. Nasal Respiration Entrains Human Limbic Oscillations and Modulates Cognitive Function. J Neurosci. 2016;36(49):12448-12467.

Zhang F, Xiao S, Kendall B, Romaniello SJ, Cui H, Meyer M, Gilleaudeau GJ, Kaufman AJ, Anbar AD. Extensive marine anoxia during the terminal Ediacaran Period. Sci Adv. 2018 Jun 20;4(6):eaan8983.

Zhulin IB, Nikolskaya AN, Galperin MY. Common extracellular sensory domains in transmembrane receptors for diverse signal transduction pathways in bacteria and archaea. J Bacteriol. 2003 Jan;185(1):285-94.

Index

Author Bio

Dr. Sauvé is a recently retired neurophysiologist from the University of Alberta. He authored 97 peer-reviewed papers on vision, developed and taught sensory physiology courses, headed Clinical Electrophysiology of Vision, directed research in Ophthalmology and Visual Sciences, was adjunct professor of Physiology and coordinated the Honours Neuroscience program. For quick access to papers: https://pubmed.ncbi.nlm.nih.gov/?term=sauve+y&sort=date

Review Ask

Don't forget to leave a review!
Every review matters, and it matters a lot!
Head over to Amazon or wherever you purchased this book to leave an honest review for me.
I thank you.

Made in the USA
Monee, IL
04 November 2026

67381649R10184